BLINDING POLYPHEMUS

THE ITALIAN LIST

BLINDING POLYPHEMUS

Geography and the Models of the World

Franco Farinelli

TRANSLATED BY CHRISTINA CHALMERS

LONDON NEW YORK CALCUTTA

Questo libro è stato tradotto grazie ad un contributo all traduzione assegnato dal Ministero degli Affari Esteri e della Cooperazione Internazionale Italiano

This book has been translated thanks to a contribution to the translation given by the Ministry of Foreign Affairs and International Cooperation of Italy

The Italian List
SERIES EDITOR: **Alberto Toscano**

Seagull Books, 2018

Originally published as Franco Farinelli, *Geografia. Un'introduzione ai modelli del mondo*
© 2003 Giulio Einaudi editore s.p.a., Turin

First published in English translation by Seagull Books, 2018
English translation © Christina Chalmers, 2018

ISBN   978 0 8574 2 378 8

British Library Cataloguing-in-Publication Data
A catalogue record for this book is available from the British Library

Typeset by Seagull Books, Calcutta, India
Printed and bound by Maple Press, York, Pennsylvania, USA

Contents

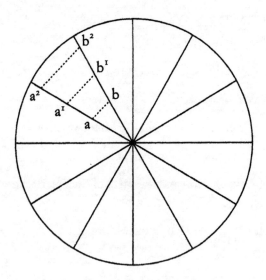

FIGURE 1. The ideal plan of the isonomic city.

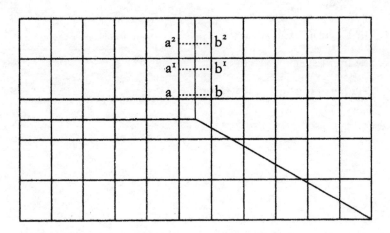

FIGURE 2. The ideal plan of the Hippodamian city.

The pre-conceived order (of geographical discourse) exists for convenience in map-making and thus proceeds from left to right, like the hand when it moves from the things it has already inscribed (on the table) onto those it must still insert. This will be done by drawing things to the north before to the south, and to the west before to the east, since our conventions have it that—for the cartographer as for the gaze of the spectator—'above' means 'north' and 'right' means 'east' of the *ecumene*, both on the globe and on the map.

Ptolemy (*Geography*, II, 1)

We turn our gaze to a globe, a minute and as ever flawed semblance of the extraordinary shape of the terrestrial sphere. The globe is capable of exercising a profound impression on our imagination and our spirit, with its perfect roundness which unites such manifold differences, and yet it is incapable of showing the smallest trace of a visible order presiding over the conflicts in the division of the waters and lands to our eyes.

Carl Ritter (1852: 206)

In general I take as true what is found in text-books, of geography for example. Why? I say: All these facts have been confirmed a hundred times over. But how do I know that? What is my evidence for it? I have a world-picture. Is it true or false? Above all it is the substratum of all my enquiring and asserting. The propositions describing it are not all equally subject to testing.

Ludwig Wittgenstein (1969: 23–4)

0. *Close an Eye, Square a Sheet of Paper: The Nature of 'Logical Space'*

Every manual, and this is in its own way a manual of geography, is founded on a double faith: that what is written about has an order and that this order is reproducible on paper, on the page. As was warned many years ago: '[O]ne cannot write a tract on a reality which cannot be factually outlined' (Maldonado 1971: 9). But a book of geography is not just any book since, more than any other, it, even if the author does not realize, refers from the outset to the whole world, that thing which, without knowing so any longer, we indicate every time that we open our arms in order to signify resignation: a gesture which refers to the impossibility of the work which we happen to have before us, but which precisely derives from the first original attempt, which is to grab and carry with oneself, in the desired direction, the 'totality of facts' (Wittgenstein 1922: 1.1) of which the world is composed. In this gesture, in fact, the limbs are not fully stretched but slightly arched, the elbow is not rigid and the fingers point forwards along the axis of the forearm, because the totality that one would like to embrace has a spherical form: it is indeed the globe, the terrestrial sphere, the ball, or rather the 'bale' of modern translators of Ptolemy (§§ 4–5),* a term that is nowadays, in current language use, already synonymous with what is fiction and fib, or else a state of Dionysiac intoxication (§ 2), a drunkenness. The impossibility in this way becomes incredulity, as though only through recourse to fantasy or the unconscious could we take account of the world as it really is.

---

* In this way the author cross-refers to chapters in this volume. [Ed.]

In order to be the world, facts must be in 'logical space' (Wittgenstein 1922: 1.13). If the philosophers were to read the geographers and vice versa, it would have long been understood that such an expression amounts to cartographic representation, to the map. It would have been understood that the *Tractatus* is the only true manual of cartographic logic already written, the most coherent attempt at geographic comprehension of the world, based, that is, on the reduction of the world to a geographic chart. When we were children, they taught us nothing. Or, rather, they have taught us to do things without any longer having memory of their meaning. Nobody has ever explained to us that calligraphic letter stems, those small, extremely artificial rectilinear segments with which we were introduced to the mystery of writing, were once the spears of warriors. Nobody has ever explained to us that every time we square a sheet of paper with line and compass we return like Ulysses to blind Polyphemus, to reduce the world to space. Polyphemus, the 'monster of illogical thought', represents the world prior to all reason, power based on pure physical force (§ 1). And this world coincides with the globe, with the enormous, heavy mass which blocks the entrance to the cave and prevents the Greeks from returning to freedom. For these men, when they finally succeed in returning to the light, nothing will truly be as before; between themselves and the world there will be something that previously there was not: the Earth.

The aggression against Polyphemus is unleashed only after the giant has stretched out on the ground, drunk on wine and human flesh, after his size has mutated from towering and vertical to a horizontal extension. In action, in this way, two axes or lines come into contact: that of the body lying on the ground and that of the wooden stake held by five trembling human beings. Staggered along the stem at regular intervals, these men constitute a true and proper living scale, an archetype and mould for that metric or graphic which still today distinguishes a cartographic representation from a simple drawing. Even today the notches along the line—which corresponds precisely to the stake, smooth and rendered straight—represent Ulysses and his comrades, exactly in the order of their attack. At one extremity the leader, and his men at identical distances one from the other. Together the body and the stake rig out two half-diagonals at right angles, centred on the crossing point at their ends: to better thrust the stake it is necessary to have a certain width of angle, and in Verse 382 of the *Odyssey* it is said that the body is 'raised', and thus we might suppose that such width does not differ much from 90 degrees. It is exactly and only because the eye serves as a central point that Polyphemus is a Cyclops. That is, he is a being of the circular eye (or face), whose surrounding appears already predisposed for its function, already ready for the traumatic insertion which signals the birth of centrality. The incandescent stake 'burns' the perimeter of the eye and 'fries' its roots, the text says. In such a way all depth is negated; there remains only a flat expanse where there was once a globe. So savagely cored and defined, the centre still scalds: within the circular assembly which delimits the first form of political activity, as in the ideal profile of the city, no warrior or citizen will be able to long occupy the position, but after a short time will have to cede it to another. The end result of this to-and-fro will be what we call 'democracy' (§ 76–79) (Vernant 1966; It. trans.: 218–69).*

But how long is the wooden stake? Ulysses commands to cut it in half at the length of two arms, according to the text: his arms—he thinks—since throughout the whole episode the stake acts as a prosthesis of his body. We are talking in this case of arms that are well stretched, on the axis from the shoulder to the fingertips, as rigid and straight as possible, a foreshadowing of rectilinear syntax (the contrary of spherical

---

* The author often refers to Italian translations of works originally composed in other languages. In the reference system, the date of the publication of the original edition is mentioned for clarity while the page numbers of the Italian translation, wherever applicable, are indicated with 'It. trans.' Details of both editions are listed in 'Works Cited'. [Ed.]

syntax: § 4) whose course will truly bring us to salvation. This kind of measure is in any case crucial, because it finally enables the development of the two lines of the body and the torso in the two diagonals we first trace when we square a sheet of paper. It also permits us to comprehend what the compass really is. To cut a stake in half at the length of two arms requires, above all, the opening of a pair of arms, with the consequent automatic intervention of the symmetry—between the right and left—which belongs to the human body. It is exactly this symmetry which governs the extension into real, full diagonals of the two original half-diagonals: the centre remains fixed, but in this way it becomes the cross-point of four half-lines, the second pair of which is the mirror image of the first, and occupies the other half of the sheet—which in such a way is fully traversed from one vertice to the other. After which, those who draw put aside their line and pencil—which are two different and distinct versions of the olivewood stake—and open their compass, which is nothing more than the two arms of Ulysses, each loaded with one of the two functions of the sharp and charred stake, which are to sting and to write. The world can finally be transformed into its own model, the introduction can commence.

PART ONE

The Pyramid and the Triangle

1. *The Two Names of the Earth*

Geography is the description of the Earth. This has been repeated for centuries. But this is not true, because, in the meantime, the most important thing has been forgotten: that it is exactly through this description that the world is reduced to the Earth, the Earth to its surface and this surface to a table. Such a definition implies, therefore, a triple transformation which even if at first passes unnoticed, later becomes uncontrollable.

The world is the complex of relations (social, economic, political, cultural) within which human life unfolds. This is the same as it was for the ancient Greeks: a hierarchy, a complex of relations of power, of relationships of authority (Vernant 1962: 114). More debatable is establishing what the Earth is, since every definition implies a personal point of view. At the beginning of the Common Era, Strabo will rebuke Eratosthenes (who, three centuries later, was the first to title a work *Geography*) to have conceived the Earth not as a geographer but as an astronomer, preoccupied, above all, to measure the Earth as though it were just any celestial body. What Strabo describes—in his 17 books of geography (I, 4, 7; II, 1, 11; II, 1, 41; II, 5, 4; II, 5, 5)—is not the world as a whole, but only that part of it which he knows and for which he possesses a language: something which in classical geography takes the name of *ecumene*, the world insofar as it is known and inhabited, and which in his case coincided in practice with the lands surrounding the Mediterranean, the Black Sea and the Red Sea.

For Carl Ritter, at the beginning of the nineteenth century, the Earth was instead 'the home of the education of humanity'. In his vision, that is, the forms of the Earth's surface (seas, mountains, deserts) represent a true project; they constitute the giant writing in which God practically directs the history of men along the path of redemption, towards salvation. The adoption of a decidedly religious perspective did not prevent Ritter, follower of Strabo, from being the most important geographer of his century, the founder of modern geography. He calls geography *Erdkunde*, a term which can be translated as 'historical-critical consciousness of the Earth', and he explains in the first of his 19 volumes that every scientific work, meaning every analysis which is objective as far as possible, is dependent on an 'ideal point of control', is built on a choice of values which is absolutely subjective and not scientific, because before being a scientist one is a human being living in society. This is what corresponds for Ritter to the 'human point of view', based on which, however, he is the first to ask of the Earth itself criteria for its own description. Still today, indeed, we view the Earth as first Ritter taught us to, as an ensemble of regions each characterized by a particular and specific complex of relations, among those which Ritter distinguished as 'geographical dimensions' and 'physical dimensions'. The first term is constituted by length and breadth and, therefore, corresponds to the plains; the second term is hinged on depth and height and thus coincides with depressions and reliefs (§ 14, 39) (Ritter 1852: 62, 25–6, 6, 27–8, 72–5). In the book which begins here, Earth is used to mean the material base, and thus the visible base, of the world.

It remains to be explained why that material base must be a surface; where, that is, its form derives from. To begin to do this we should remember the text, composed in the first half of the thirteenth century, which is known as the first geography of the Earth in the Spanish language: *La semejanza del mundo*, or *The World's Resemblance*. The title depends on the fact that the work is conceived as a kind of mirror of the world (Bull and Williams 1959), and a mirror reflects only what the philosophers define as the phenomenal aspect of things, or what one immediately sees. It should be noted, in this regard, that the first of the two Greek terms which the phrase 'geography' is composed of, *Gé*, means Gaia in Latin, and thus the Land which glimmers and shines in the light. The other name, the first, with which the Greeks indicated the Earth, was *Chthon*, which in Italian survives only in the adjective 'ctonico' (English, 'chthonic') which means subterranean, cavernous: a name which resounds, asks to be heard, exactly like the environment it refers to.

2. What Is Geography and Who (What) Is Dionysus?

Between *Gé* and *Chthon* there is a systematic opposition: the first refers to the Earth as something evident and, therefore, clear, superficial, disposed according to horizontal movement. Contrary to this, the second implies invisibility, darkness, the interior and not the exterior, depth and verticality and not horizontality. Geography is the description which corresponds to the first mode which is precisely that of mirror vision. This does not mean that it is the only mode possible, and even less that it is the oldest we can remember. We pay for it, it comes at a price. A myth narrates the origin of this mode, which is the myth of the killing of Dionysus (son of Zeus and Persephone, and also the subterranean deity) by the Titans, the sons of Chthon himself. For this purpose, the Titans stained their faces white with chalky powder, and then scattered chalk on the face of the divine sleeping child. When he, awakened, looks at himself in the mirror, astounded not to recognize what he sees, he does not recognize himself. Right at this moment of the god's astonishment, of the fixity of his gaze on something unexpected and which he has never seen before, the Titans take advantage, to kill him and to cut his body into seven pieces. As will be explained centuries and centuries later by the anonymous compiler of the *Resemblance*: the world has the form of a ball or an egg, just like a person's head, and the problem of knowledge consists in deconstructing it into its elements, subdividing it into parts (ibid.: 53).

As in all of Antiquity (just think of Cicero and Seneca), so it was in the medieval period: it was not, in fact, believed that the Earth was round. It was well known that the Earth was spherical, contrary to what (on the subject of the so-called Dark Ages) began to be believed from the beginning of the nineteenth century (Burton Russell 1991). But this is not the point. And neither should we linger too long on the fact that it is exactly in this 'making the world into pieces' that the philosophy of thinkers like Ludwig Wittgenstein consisted. On this point, Strabo (I, 1, 1) is very clear, from the very first line of the first book of his writings: starting from Homer, and practically until Aristotle, all those that wrote anything were geographers, and in particular those that we still call pre-Socratic philosophers—for Giorgio Colli (1977–80), they were 'Greek scholars'. In other terms: philosophy is a development of geography; it is born from it, and from it—the original form of Western knowledge—it assumes its models and figures of thought. But as the myth teaches, all begins when instead of Dionysus, the god of uninterrupted and limitless life, of life understood as an infinite and indistinguishable (thus inseparable) process (Kerényi 1976), the mirror reflects the white veil of land which covers his face and hides it from his eyes. It reflects, that is, his face transformed into a clear surface which, exactly because it is for the first time distinguishable, has never been seen. Only through the effect of such a transformation-substitution can the swords and knives of the Titans start operating and section off the totality of vital process, taking advantage of the moment which corresponds to its partial paralysis. And only with such blades is it possible to obtain boundaries, limits, lines which separate and define things, which section and divide them, rendering thus possible our life which, precisely by virtue of these limitations, is different from that of the gods.

Dionysus, the oscillating and swinging god, the god who vacillates, is thus the globe, the world. The chalk is the Earth reduced to surface (*Gé* itself, from which the term derives) and the blades are our concepts, more or less sharpened. In this inventory of the elements of the sacrifice which births Western knowledge, however, we are lacking one element, the most important and elusive because the most common. And indeed there is no version of the myth which stresses it. It is only said that Dionysus returns to life because his brother Apollo, god of measure, recomposes his body at the behest of Zeus. It is not possible, however, to put the limbs back together again without leaning them on a surface which in this way becomes the first altar: a table that, as in all cartographic representation, serves only two of its dimensions, length and breadth, since it is as flat as possible. Much like the mirror, which at the beginning of the story reflects the clarity and superficiality of something which is still whole; while the altar is the table which, in the end, imposes horizontality and contains and recomposes the dismembered whole.

3. *The Book of Islands and the Atlas, Place and Space*

It would however be naive to think that the table serves only passively to welcome what remains of the globe. It instead transforms the globe in crucial respects, along with our manner of entering into relation with it. On the table the pieces remain as pieces; at the same time they constitute a unity. This is possible by virtue of the lines which distinguish them and at the same time unite them, but that only manifest on the table, and which, therefore, are a product of the table; they originate from it as well as from the cut of the blades. Recomposition consists in the juxtaposition of pieces, in the placing of them one next to the other in accordance with the original model. In such a way, even if it does not seem so, the nature and functioning of the globe emerge radically modified. Dionysus, the myth recounts, made Arianna pregnant, and Dionysus was born from Arianna. In the same mode as Hindu mythology, for example: Mount Meru, the axis that supports the world, has its base in the Himalayas. For the myth, in short, things exist one inside the other, like Russian dolls or the layers of an onion, and for this reason we have trouble distinguishing them. The physics which results appears ridiculous to us, like the baron of Munchausen's claim to have risen out of his chair by grabbing himself by the hair and pulling hard. In the language of cybernetics, the motion by which things lurk on the inside of other things is very important, and is called recursion (Hofstädter 1979; It. trans.: 137–47).

As in the myth, on the globe, things are arranged according to a relation of recursion, and until the end of the sixteenth century, geographic description also obeyed this relation, in some respect. Today we normally subdivide the world into continents, meaning into large, continuous, definite masses of land, ideally separated by oceans. Anglo-Saxon usage individuates, in order of volume, seven continents exactly, like the pieces of Dionysus' body: Asia, Africa, North America, South America, the Antarctic, Europe and Oceania. There exist subdivisions that are different from these, where, for example, Europe and Asia, which are in effect not separated by the sea, form Eurasia. 'Continent' is a term which signifies something that contains something else but, despite its meaning, it does not recall any recursive conception. It begins to emerge between the fifth and the sixth centuries and is definitively imposed in the nineteenth century (Lewis and Wigen 1997: 28–31), following the wider and wider circulation and spread of atlases. The first collection of maps with a frontispiece decorated with Atlas holding up the globe was printed in 1570 in Rome by Antonio Lafreri. Before the atlas there were only isolariums, books composed of maps and descriptions in which the entire globe, beginning with the Mediterranean, was decomposed into islands, into something, that is, which before containing something was instead, by definition, contained in something else—the sea. Islands were considered to be all the existing lands, from the very small ones to those extremely large ones recently discovered in the Western ocean (the 'land of Santa Croce', as America came to be named in the 1528 *Isolario* of Benedetto Bordone). There is one essential difference between the atlas and the isolarium: in the first, the globe becomes transformed into space; in the second, on the contrary, such transformation does not occur, and the existing lands are still considered as places.

Space, it must clarified at this point, is a word which derives from the Greek *stadion*. For the ancient Greeks, the stadium was the unit for measuring distances, and thus signified literally a standard linear metrical interval. Deriving from it is the fact that, within space, all the parts are equivalent to one another, in the sense that they submit to the same abstract rule which takes no account of their qualitative differences. Such a rule is represented by the scale, which from the sixteenth century begins to appear systematically on maps (P. D. A. Harvey 1985), and indicates the relationship between linear distances in drawing and those that exist in reality. Place, on the contrary, is a part of the Earth's surface which does not match any other and which cannot be exchanged with any other without everything changing (§ 58). In space, instead, every part can be substituted by another without anything being altered, exactly as when two things which have the same weight are moved from one side to the other of a set of scales without the equilibrium being compromised.

4. *The Birth of Space*

To say equivalent in Greek, you say 'parallel': the invention of space is exactly thanks to the introduction, in the description of the Earth, of what is called the 'geographical grid', which is to say, the network of meridians and parallels with which one seeks to reproduce the curvature of the globe on paper. Such a process of restitution is called, in modern terminology, 'projection', a word which derives from alchemy (Eco 1990: 76) and which refers to the most extraordinary transformation, that of crude metal into gold, precisely ensured by the powder of projection. Cartographic projection is based on a mathematical rule which allows one to determine the correspondence—on the flat surface of paper—of one and one point only with any given point on the globe by the intersection of a meridian with a parallel (Fiorini 1881; Snyder 1993). In other and more immediate terms, it is equivalent to a truly formidable metamorphosis, which is to transform, in a coherent manner, something which has three dimensions into something which has two by subtracting one dimension of the Earth. Such subtraction is rendered necessary because the sphere and the plane are irreducible to one another, since their surfaces do not have the same properties: the first is round and finite, the second conversely open, its lines not at all closed (Reichenbach 1957: 59). Consequently, only the second, meaning cartographic representation, allows infinite process and uninterrupted expansion which, in all respects, characterize the modern epoch and Western culture in particular.

The first to confront this problem of cartographic projection seems to have been Erastothenes, in the third century BCE (Prontera 1997). But it was the *Geography* of the Egyptian Ptolemy, written in Greek in the second century CE, which would transmit to the modern era the method of transforming the Earth into space, the sphere into the map. Already for Ptolemy, the geographer of the Roman Empire at the height of its splendour, the Earth was a head, which he teaches one to reduce to a plane according to systems of projection much more sophisticated and precise than all the preceding systems. With the fall of the empire, his work disappeared almost entirely from the horizon of Western culture, to reappear after a thousand years thanks to the influence of Byzantine culture. Art historians (Edgerton 1975; Veltman 1980) are convinced that the invention of modern perspective (linear, Florentine perspective) is a direct consequence of this reappearance which occurred in Florence exactly at the beginning of the fifteenth century. In effect, Ptolemaic projection and linear perspective are the same thing. Both presuppose a fixed and immobile subject; both reduce knowledge to vision, to a matter linked exclusively to the eye, and which is, therefore, instantaneous. Both conceive of the order of things on the plane as depending upon the simple distance between them. Modern perspective also makes the dimension of objects depend on distance, different from the perspective of the Ancients, for which the dimension of objects was determined (much more correctly) by the width of the visual angle (Panofsky 1927). But perspective and projection also have another characteristic in common, which lies at the root of modern territoriality. Both represent what can be seen within a field which is imbued with the same properties that traditional (Euclidean) geometry generally assigns to extension. These are: continuity, that is, the absence of interruptions; homogeneity, that is, the identity of the material which it is composed of; isotropy, that is, the equivalence of the parts with respect to direction. Such properties are, in all evidence, those which concretely belong to the table, the map, which means to the material support of geographic representation (Farinelli 1997: 43–59). Together with the attribute of measurability, these reflect the characteristics of the altar on which Apollo reassembles the body of Dionysus exhaustively and to perfection. Hence, in the five centuries which will pass between Erastothenes and Ptolemy, these characteristics will extend into the image of the pieces of the world, to cartographic representation itself. Apollo is, in fact, the executor of the first exemplar of what we today call a map, for which his brother will involuntarily furnish the raw material, and the altar provide the rules and characteristics. It is precisely this last element, the altar, which will establish the model of space which corresponds to extension in Euclidean geometry.

5. 'The Age of the World Picture'

Until Ptolemy, geographic representation, that is, the reduction of the world to a table, is concerned only with the things one sees. With modern perspective, it instead invests and colonizes also what is not seen: the interval and gaping void between the subject looking at the world and the object looked at. The eye which runs over this distance can no longer be stopped during its course but must instead look at everything at once, magnetized by the vanishing point behind which infinite empty space hides. And before which, instead, stretches the mathematical order of full finite space, the 'space filled with earthly things'—as described by the German geographers of the early nineteenth century—in which everything is transformed into measure, continuity, homogeneity, isotropy. Still before this, however, the subject is reduced to the eye which, as Leon Battista Alberti explains, becomes the prince of the senses, the only organ qualified for knowledge. A flying eye will, therefore, become the emblem which Alberti chooses for himself (Smith 1994). All the rest of the body is thus transformed into a pair of wings, and such a metamorphosis reflects better than anything else the intention which animates the transformation of the Earth's surface into space: the reduction of travel time, the increase in the speed of movement of men and goods from one point to another on the globe. This is nothing new in relation to the logic of the Roman Empire, to the *celeritas* of Julius Caesar (Rambaud 1974). His famous motto *Veni, vidi, vici* means exactly this: I have won because I have reduced knowledge to vision (*vidi*) and I have done this before others (*veni*).

All the great empires of the past translated themselves into grand road systems, tending to be straight in order to be as fast as possible: from the Chinese Empire to that of the Inca, from the Napoleonic system to the English one. But the straight roads which were the pride and advantage of the Roman Empire were subject to the same fate, of abandonment and forgetting, as Ptolemy's *Geography*. Nonetheless, even the syntax of the modern territory is constituted principally through the rectilinear. It is precisely perspective, and thus projection, which functions as a vehicle for the reintroduction of the rectilinear model of the world's functioning, for the spreading and generalization of what in the past was the imperial model. This is not merely an immaterial model, an impalpable one, which becomes material and suppresses everything based on the curved line. It is also an extremely pervasive model: in its uniqueness, it serves at once to perceive, to represent and to construct the face of the Earth, so as to colonize all the forms of our relationship with it (Farinelli 1986). Almost until the end of the seventeenth century, however strange it may seem, maps which showed the layout of land routes were truly few (Dainville 1964: 261–3): the form of the road took the form of courses of water as its model, since the two were not infrequently parallel. Thus the cartographer, not able to represent everything, represented the courses of water, and not the land routes which were often less important. But starting from the eighteenth century, roads, precisely because they were straight, were made separate from the structure of the water system, and began as such to stand out also on maps because they could no longer be assimilated to the sinuous waterways. One would do well to find a more immediate and concrete example of the difference between the pre-modern and the modern period. At the root of this difference lies the complete overturning of the relation between cartographic image and reality. In the medieval period, cartographic representations were the copy of the world; they mirrored the relations of which the world was composed, and were thus a religious and philosophical interpretation of it as well as a design of it (Edson 1997). They were its portrait but also, consciously, the self-portrait of the culture which produced the portrait (Barber 2001). Conversely for Heidegger (1950: 71–101), who is, for many, the most important philosopher of the twentieth century, of modernity and of the 'Age of the World Picture', the first modern movement consists in the reduction of the world to an image, or, insofar as we are concerned, to a map. This is as much as to say that for the modern epoch, opposite to the medieval period, it is not the map which is the copy of the world but the world which is the copy of the map. It is in this way that the world is truly transformed into the surface of the Earth (Farinelli 1989a).

6. *The World's Duration: Marco Polo*

As on isolariums, space was rare in the medieval period and the world was often made up of a set of places. Every place has its own measure, in such a way that none of them is standard. The things of the world are limited in that they must remain within these proportions, as on the globe, for which there is no scale, and on which thus there is not, strictly speaking, even a tiny piece of space. Consequently, in the medieval period, unless one is a messenger or a soldier, the problem of speed in general does not exist. This is true also for merchants whose weapon is the secrecy of relations rather than their speed. Take the case of Marco Polo, the most famous of medieval merchants and travellers, who in the last quarter of the thirteenth century arrived in China from Venice along the 'Silk Road' through Persia, Afghanistan and Turkestan. He rode across a long and perilous road, even if it had been known for millennia, and every day he was presented with the opportunity, if not the necessity, of changing his route, as well as of stopping. In the Chinese city of Ganzhou (today Zhangye, past the desert of Taklamakan and west of the Yellow River), Marco, together with his father Niccolò and his uncle Matteo, stayed an entire year, at their own pleasure. We understand, then, how Marco knew all the idioms of the countries he had crossed: the Turkish spoken by Mongolians, Arabized Persian, Mongolian, the Turkish spoken by the Uighurs of Xinjiang. Before turning back, the Polos lived in the domains of the Grand Khan, the emperor of the Mongolians, for almost 17 years.

Marco thus rides without hurry, stopping every evening in the caravanserai and for whole months, if necessary or for pleasure, in the cities, learning languages and customs, information and stories. Every day the things of the world reveal their proper duration to him, and at the same time measure the duration of his life. Indeed, in the *Travels*, the extraordinary account of the travels of Marco, the deserts, forests and mountains do not already have an extent, just as the directions of the route are not already fixed according to the abstract rigidity of compass points. To advance, he takes towards the north or north-east, following the directions of the winds, following their course. In this vein we can read, in the fourteenth-century Ottimo version of the *Travels*, expressions of the kind: 'Carcam is a province which lasts for five days' or 'When man takes off and has ridden these 20 days of the Cuncum mountains', etc. It is as though in the *Travels* space does not exist, in the same way that time does not exist, if not in the form of the alternating of the night and the day and of the seasons. On the contrary, places and days are the same thing: they coincide in the experience of the journey, and the first serve as a measure of the second and vice versa. This means a relative measure which changes in turn, and which has nothing of the metrical, linear or standard in it. Like the places, the days are also not uniform. Meanwhile, climatic conditions vary continually: for their return to China, the Polos will take three and a half years, thanks to the snow, rain and heavy floods, and because riding in the winter is an entirely different matter from riding in the summer. Furthermore, the nature of places constantly changes and, consequently, so does the means of locomotion. From the fourteenth-century sources, we derive that the normal time it takes to arrive in China from Tana in the Crimea was, then, for a merchant, around 9 months, in the order thus subdivided: 25 days with oxcarts, 9 by water, 50 with a camel train, 115 with pack donkeys, 75 on horseback (Larner 1999: 187–90).

There existed just one alternative, thanks to which it took up to a tenth of the normal time: the *yam*, the postal system of the Mongolian Empire, based on a chain of stations for messengers who from the capital Cambaluc branched off over the whole kingdom at intervals of 25 miles from one another. It is the sole example of space which Marco describes, a domain of linearity and thus of speed and the equivalence of parts. But it is certainly not his world, if he ever had one. Certainly he would have remembered many more things, we read in an unedited manuscript of the *Travels*, if one day he had ever thought of turning back. Only space, which is uniform and continuous, implies return, the reversibility of movement. But if the world is composed of places, of parts which are not continuous, not homogeneous, not isotropic, it is not certain that the return will occur. On the contrary.

7. *Odysseus in Space*: *Christopher Columbus*

In the case of Christopher Columbus, the first of the modern travellers, the opposite is instead true. His problem is haste, turning back as quickly as possible. This is why he looks for the Levant by travelling towards the west. We read of the first voyage in what remains of his logbook, under the date 22 January:

> Then proceed north, quarter-north-east for the space of 6 inners, which will have been another 18 miles. Then 4 inners during the second guard to north-east at 6 miles per hour which will make 3 leagues north-east. Then until the dawn proceed to east-north-east, for 11 inners, 6 leagues every hour, which make 7 leagues. Then to east-north-east, until at 11 hours of the day, 32 miles (Varela 1982; It. trans.: 119–20).

It suffices to clarify here that every inner [*ampolletta*] is an hourglass which lasts for half an hour, and that for Columbus a league means 4 miles, to realize that these are false calculations. But they are, precisely, calculations, relations that are of spatial and temporal magnitude which are wholly conventional (the hour, the league, the mile), justified only by the fact that what is pressing is speed, meaning the abstract relation between abstract quantities. Such relations situate themselves within a framework which is just as abstract, no longer defined by the names of the winds, which nonetheless continue to blow, but by the invariable geometry of compass points. This occurs because geographic representation has already taken over the world, space has already encompassed and absorbed all places, the map already substitutes that which it represents to the point of anticipating its nature and features, and prefiguring its very existence. Consider what is recorded on the date 25 September. Both Columbus and Martín Pinzón, the commander of the *Pinta*, are by now convinced of being close to land. Such conviction is founded on the simple fact that both have 'found certain islands painted in those waters', realistically drawn on the map of the ocean prepared for Fernando Martinez, cleric of Lisbon, by Paolo dal Pozzo Toscanelli, the greatest and most mysterious among modern cosmographers, and from these transmitted in copy to the Genovese navigator. At sunset, Martín, high up at the stern, 'called most joyfully to the Admiral, asking for a reward since he had sighted land'. After which 'all climbed the mast and rigging and all confirmed the discovery of land. To the Admiral it seemed so too and that it was not more than 25 leagues away.' Only the next day 'they discovered that what they had taken for land was not land but sky' (ibid.: 19–20).

It will be said that it was a matter of simple impatience, and that, in any case, they were not very far from the coast, having by this point travelled across three quarters of the distance. The fact remains that, once arrived, they are convinced of being where they are not. Only towards the end of their days, in the course of the fourth expedition, will Columbus finally be struck by the suspicion that the land touched upon by himself is not the magnificent Cathay of Marco Polo but 'another world', a 'new world', terms which significantly begin to appear only in the account of the third voyage. If it were not fundamentally tragic, the series of misunderstandings which follows would be, as in authentic tragedies, truly hilarious at times. What in any case remains moving is the effort of Columbus, truly arrived in sight of land, to make that which he sees, and which Toscanelli has never seen, coincide with the traits and features painted onto the map which he carries with him and in which he believes blindly. In other words: in order to make the Earth conform with his cartographic image, he bashes the world into shape. If in the world of Marco Polo, where neither space nor time exists, things endure, in the world of Columbus, dominated instead by the spatio-temporal abstraction, things are, on the contrary, spread out: the mines of Veragua, he explains in the account of his last voyage, 'extend the space of 20 days to the west and are found at an equal distance from the pole and the equinoctial line' (ibid.: 395, 345–6). Here, space means, as in Ptolemy, the interval between one node and the next in the grid of meridians and parallels, presupposing projection, and, therefore, the map, not the globe. Things are exactly the opposite of what is often still believed today: the impact of Columbus was not by any means that of making the image of the Earth spherical when it was previously believed to be flat, but of transforming the whole Earth, from the sphere that it had been and had been believed to be, into a gigantic table (§ 72).

8. *The Dogfish's Tooth and the Centre of the Labyrinth*

To summarize. We only truly leave myth through projection, meaning through space, which transforms something which we do not manage to define, which it is impossible to ascertain the being of, into something whose nature and identity we instead control. History begins with the pitiable work of Apollo who reassembles the parts of his brother, and is thus constrained to act out what the table at the altar dictates to him: cartographic logic. He places the pieces one next to the other, in doing so, following an order of closeness and farness (thus, distance) between these pieces, whereas in their original state they were each organically connected to the other in such a way as to be at most distinguishable but not separate. It should be noted that this is precisely the manner in which the difference between geography and geology is determined. Geology, which studies the structure and evolution of the Earth's crust, is born by making itself autonomous from geography. It makes itself autonomous when, just past the middle of the seventeenth century, the Danish scientist and theologian Nicola Steno, or 'Stenone', poses himself a question that geographers can by this point absolutely no longer ask: How can, in Malta, the tooth of a dogfish find itself inside a rock stratum? The question is doubly forbidden for geographers, not because it concerns sharks but because, regarding the matter of a fossil, it assumes subterranean, chthonic dimensions, and, even more, the logic of recursivity. The title of the essay with which Stenone responds in 1669 shows this immediately: *De solido intra naturaliter contento dissertationis* which in English translates to 'Dissertation Concerning a Solid Body Enclosed by Process of Nature within Another Solid'. Contrary to this, as has already been demonstrated (§§ 3–5), between the sixteenth and seventeenth centuries, every recursive model disappears definitively from the spatial image to which geography reduces the world. More and more precise and mathematically reliable maps are substituted for the world. On these maps, indeed, things are ordered according to the metric scale, and, therefore, can only exist closer or farther away from the next thing, never one inside another. It is through these images that, returning to the language of Heidegger (§ 5), truth transforms itself into certainty of representation, meaning into science, whose rigour derives directly from the cadaverous rigidity of the body of Dionysus laid out on the table. The first to denounce Stenone will be Carl Ritter himself (1852: 34–5): maps are to the essence of the world much as the anatomy of the body is to the living substance of the heart. This is exactly what Ritter argued.

It remains to be shown how it is the logic of the table to proclaim law. For this, the figure of the labyrinth suffices, the image which corresponds to the cataclysm of the first projection, and to the losing of one's way that this produces. The origin of the labyrinth is actually, outside of all rhetoric, simple. It is what results from the collapse of the tower of Babel, from the Babylonian Ziggurat, from the Minoan-Cretan palace, from the Egyptian pyramid: in short, from the fall, from the crushing to the ground of every vertical structure whose different levels correspond to a hierarchical system of power. In other words: it is the product of the transformation of the world into the Earth. In this way, the different levels, i.e. the levels of power, are altered to become entirely other horizontal dimensions which are recursively disposed one inside the other, and it is precisely such recursivity which impedes the possibility of speaking of space (Farinelli 2002: 227–8). It is Theseus who will convert space into the labyrinth, measuring it with Ariadne's thread and thus managing to find the centre. But where does this centre derive from, and what produces it? The centre does not possess Theseus, nor does it belong to the structure which falls. Rather, it is incorporated into the table, into the earthly surface which receives the ruins of the collapse and thus reconfigures the tri-dimensionality of the original construction in two dimensions. We understand finally, then, the true reason for terror and, therefore, for the refusal which the labyrinth inspires in a culture, such as Western culture, in which knowledge passes from necessity towards representation. One cannot represent the labyrinth, one can only think it: representing it in whatever form means transforming it into its opposite, into something which possesses a centre (§§ 97–8).

In the case of the labyrinth, and only in this case, thought and representation become irreconcilable. This would matter little were it not also that the surface of the globe is a labyrinth, in the sense that, depending on how one moves around the sphere, any of its points can be the centre.

9. *The Line of Desire*

Like the centre, straight lines also derive from the existence of the table; they are the result of cartographic logic which is mirrored on the surface of the Earth, and in the Earth's configuration into its own image and semblance. At the beginning of the twentieth century, Jean Perrin (1948; It. trans.: 33–6), in illustrating the foundations of atomic physics, perfectly explained the difference between reality and its cartographic representation. In the latter, which is a conventional design, every curved line possesses a tangent, but only because this is part of the properties of the construction from the outset. According to the same principle, tracing a curve seems to demonstrate that every continuous function requires a derivative. Consider instead a random tract of coast. For every one of its points, it would be possible to find, on the map, a tangent, even if this varies with the varying of the scale. In reality, on the other hand, from whatever distance one looks at a coastline, it is extremely difficult to fix any kind of tangent on a point: as you approach it, that which initially seemed to be the tangent is transformed into a perpendicular line, or an oblique line, in relation to the contour. This is because in the contour we discover new irregularities at every step forwards. But we know this already: like perspective, the cartographic image, which is the product of projection, functions only because it immobilizes the subject of knowledge. The subject in this way becomes the copy of the corpse of Dionysus, as, in fact, Leon Battista Alberti first recognizes, detaching the eye from the rest of the body (§ 5). If the foot follows the eye and approaches the coast, the tangent is no longer a tangent. This occurs since only a single step is needed and the coast, different from before, no longer seems straight at the point where the imaginary tangent passed, because straight lines do not exist in nature. On another note, and even more importantly: Why do we consider the same coast as a line, and not a strip, which is what it is in reality? This is exactly the question which Friedrich Ratzel, the last heir of the *Erdkunde*, poses at the end of the nineteenth century (1899; It. trans. p. 284). The response is: Because we exchange reality with its cartographic image.

However, straight lines do exist on the face of the Earth, and they are themselves the proof that the Earth is the copy of the map. The same year in which Stenone writes his dissertation, in France, Jean Picard begins the construction of the Paris meridian, at the behest of Louis XIV and under the command of the minister Colbert: from Dunkirk to Perpignan, and so through the whole length of France, a gigantic line is traced on the ground, in order to be able to accurately calculate the radius of the Earth's sphere. This work, completed in 1720 by Giacomo Cassini, would have been enough on its own, according to Voltaire, to make the century of the Sun King eternal (Francheville 1752: 161). For the first time, a line in the abstract geographic grid became material, the Earth was modelled in the form of its own drawing, it became the copy of the copy itself. Such a copy becomes the concrete model of rectilinear organization in the modern territory. Already, before the French meridian was even finished, the economist Orry revealed its true function: to serve as the basis and model for the straightening of all French roads (§ 67), for the transformation of all routes into straight axes (Vayssière 1980: 254–5, 257). In the eighteenth century, this will take place in almost all of Europe, in terms of the roads (§ 5). In the nineteenth century, there will be the rise of the railways, straighter than all existing routes. In the twentieth century, motorways will be even straighter, or at least faster, because they will allow one to avoid crossing the city. At the beginning of the 1960s, the Buchanan report on the conditions of traffic in cities in the British Isles will describe the rectilinear paths which every driver mentally anticipates in order to move from one point to the next in the metropolis as 'lines of desire' (1963: 52, 250–1). These are ideal lines: they are possible on the theoretical plane and are almost never practicable due to restrictions and one-way streets; therefore, they do not actually exist. The straight line which the eye, at the outset of the modern period, follows in the perspectival gaze is virtual, in the sense that it already exists but it is not already concrete, as in, actual. It becomes actual in the construction of the modern territory, which constitutes its keystone. And in the last half century, it becomes virtual again; it is no longer actual, since it is no longer part of experience; one can only think it.

10. *The Immutable Metre and the Eternal Triangle*

In the preface to the sixth book of his *oeuvre*, dedicated to architecture, Vitruvius, the most famous architect of ancient Rome, tells the story of Aristippus, the Socratic philosopher. Shipwrecked on the deserted coast of the island of Rhodes and not knowing where he had landed, he understood from geometric figures traced in the sand that he had happened upon a place inhabited by civilized people. Without the table or a flat surface which resembles it, geometry, which for the Ancients was equivalent to civilization, would not exist. All of modern science has followed Galileo Galilei in his conviction that the book of nature 'is written in mathematical language, and the letters are triangles, circles, and other geometric figures' (1965[1623]: 38). For Michel Serres, *geo-metry* means that the metre (the measure) is the Earth (1993: It. trans.: 269). In the same sense, *geo-graphy* has almost always meant, then, that writing (geo-metric writing) is the Earth. For this reason, Ritter called geography by another name (§ 1). He was not at all in agreement with the reduction of the Earth to the deadly table—mirror and altar all at once—which is the map, and he protested with all his force against what he termed 'cartographic dictatorship' (1852: 33) in terms of geographic description. The geometric writing of the Earth is, in fact, nothing other than its cartographic representation.

Measure becomes the Earth and, vice versa, the Earth becomes measure precisely with the French meridian. In 1791, the French National Assembly turned to the Academy of Sciences to find a remedy for the confusion generated by the fact that every region or district of the country functioned using a different system of weights and measures. The 'Declaration of the Rights of Man and of the Citizen' had already made clear, immediately after the storming of the Bastille, that the acts of the assembly had a universal value; they concerned, that is, the whole of humanity and not solely France. It was necessary, therefore, to find a unity of measure which was stable and immutable and which could potentially have value for the entire globe. But not only this. Since humanity is the greatest subject that one can imagine and is composed of all persons who have lived, of all those who live and will live, such unity had to have value for all future time too. Thus came the metre as we know it today. This is equivalent to the 40-millionth part of the circumference of the Earth, calculated starting from the first meridian, precisely that which passes through Paris. Such an operation illustrates better than anything else, since it anticipates it, the destiny of the French Revolution and the ideals of the Republicans which were affirmed by it. Even these will be destined, in the following centuries, for a great extension to the global scale, exactly as the calculations which the metre derived from consisted in the extension of local values onto the surface of the globe, and so, their transformation into general values. All of this is born from the French segment of the first major, concrete meridian. But this meridian, axis of the first modern map, the map of France whose survey comes to a close right on the eve of the revolution (Gallois 1999), owes its definition to the procedure of triangulation, the only capable of assuring the accuracy and geometric precision of drawing.

One of the last inventors of non-Euclidean geometries, Benoit Mandelbrot, underlined sometime ago mathematicians' lack of memory, and the paradoxical situation deriving from it (1987: 19–20). Mathematicians are so enamoured of the models they use as to no longer remember that they are only models; they exchange them, thus, with reality. So we continue to consider the continuity of objects in Euclidean geometry as something given a priori, in the same manner as, for example, the continuity of curves in calculus. But this, it should be added, happens because Euclidean models have not served solely to describe the world but literally to construct it, to configure it; they have thus themselves become concrete reality. Properly understood, all of cartography serves no other purpose than this, to transform the invisible into the visible, software into hardware, what you can draw into what you can touch, even if we are used to believing precisely the opposite. The triangle, for example, is not only the model of a shape, but also the model of a productive process. As indeed the nature of triangulation demonstrates.

11. *The Cartographic Triangle*

Although triangulation is a very ancient procedure, already known to the Egyptians and the ancient Greeks, it is said that the first to make use of it was Leon Battista Alberti, probably because in his *Mathematical Games*, written around 1445, he provides its first modern description (Vagnetti 1980). In reality, he does something more, since his winged eye (§§ 5, 9), detached from the rest of the body and launched on ahead, declares with clarity and immediacy the assumption which governs its functioning. Triangulation puts one of the fundamental laws of plane trigonometry to use, according to which if one side and two angles of a triangle are known, one can easily derive the other angle and the length of the remaining two sides. It is sufficient to establish accurately the length of a segment, pick out an external point which allows one to measure the angles with the edges of the segment in question, and without much effort, and by way of simple calculation (thus in an abstract manner), one can obtain all the other values. All of the body is mobilized, in its entirety, only in the measurement of the first section, travelled across step by step with measuring tape in hand, and in the ups and downs of the towers, steeples, elevated points from which to proceed to indirect measurement, and which coincide materially with the three vertices of the figure. All the rest of the work falls to the protractor, compass, pencil and piece of paper handled by the same observer (become, however, immobile). The observer's eye, not simply part but substitute for the whole body, is the sole organ which crosses the sides of the triangle which are opposite to the base directly measured. The advantage of the operation does not consist solely in the economy of time and in the precision of the result but also in the possibility of calculating angles and distances which are otherwise (in concrete terms) impossible to measure. And through adjacent triangles, constructed, that is, one starting from the other, a single base is sufficient to govern the grid needed to cover an entire region.

The limit of this reduction of the world to a series of triangles consisted in the fact that every grid was auto-referential, in the sense that the principle which assured its precision and coherence was held internally. In this way, two distinct grids were never perfectly adjacent to each other, since each was turned in a different direction, directed towards a different system of reference. Each was founded on a different continuity, on a different homogeneity, on a distinct isotropy. To juxtapose on the same map two grids referring to adjacent regions meant inevitably to discover their reciprocal irreducibility. The relative difference in orientation was also normally added to the difference between the bases. This difference in orientation often did not take magnetic variation even minimally into account. Only in the seventeenth century (Pouls 1980) did another method begin to be practised, which is to calculate distances on the Earth no longer through picking out the highest objects in a single region but by making reference to angular distances between the stars and the planets. Thus the system of celestial bodies took over as a system of correspondence, which, being external to the Earth, assured an even greater precision, and, above all, presented the advantage of being common to an entire hemisphere and no longer only local. In this way, it was possible to proceed to the construction of cartographic representation of regions which were much larger, as, in fact, centralized territorial states were as compared to previous political formations. In such a way, as the story of the French meridian shows (§§ 9–10), the definitive projection onto the face of the Earth of an order identical in nature to the order which defined the perception of celestial bodies was made possible: geometric order. It was precisely this projection that transformed the Earth into the modern territory.

The French example was the first instance of triangulation on the state level. Thanks to its success, it was adopted in the nineteenth century in all of Europe and in many colonized countries. But the importance of triangulation was not limited to the construction of the cartographic image of the territory, and thus of the territory itself. As a constructive procedure, it also constituted, just like the perspective from which it derived, an extremely powerful and efficient model of knowledge, whose effects have dominated the following century and even today retain control over our relationship with the world.

12. *The Semiotic Triangle*

Triangulation functions through the substitution of step by sight, and it is exactly this substitution which is the basis of the modern form of passage from symbol to sign. For the ancient Greeks, the symbol was an object split in halves and containing two different people who, meeting each other and returning to match the separate pieces, guaranteed, reciprocally, the identity of the other. It was, then, a system of recognition but, at least in origin, not a mere sign, since each fragment represented a concrete relation, for example, of knowledge or friendship: the descendant of a family who had offered hospitality to someone, making a visit in turn, after years, to the town of his guest, and bringing with him the fragment which testified to the past relationship, assuring him the hospitality on the part of the descendants of the guest in question. The sight of the symbol, in cases like this, was not a substitute for the step but was instead the result and the point of arrival of all the steps which made up the journey of the traveller, plus those of the ancestor who had begun, with his own voyage, the link between the two families. The sight of the symbol was thus something which stood for the total of steps effectively made from the beginning of the affair, and only because it represented them (that is, it returned to make them present) was it capable of substituting them.

Even the sign is something which stands in for something else, which refers to something (or someone) that is absent, and, therefore, it also presupposes a distance. But different from what takes place for the symbol, Hegel explained (1955; It. trans.: 402–03), the connection between expression and meaning is wholly arbitrary in the sign, in the sense that it is external and formal. That is to say, returning to our example, that between the sight of the sign and the steps, there is no relation whatsoever, and the sign substitutes steps without any longer representing them. The sign is no longer the product of the voyage; on the contrary, it dispenses with the journey, it renders it superfluous. Just as happens only in triangulation, so with the cartographic sign. For the semioticians, the world consists of an endless universe of signs. And it is not surprising that it is their custom to resort to a triangular graphic in order to synthesize their own cognitive model. The first to make use of such a device were, in 1923, C. K. Ogden and I. A. Richards in their volume on *The Meaning of Meaning* (see Eco 1989: *x*). The upper vertice of the triangle generally comes to correspond to any given *signified*; the lower vertices are arranged on one side as the *signifier*, for example, the word which expresses the signified in any specific language, let's say 'bicycle'; and on the other side, the *referent*, or the object bicycle here before me or else all the bicycles which exist, have existed and will exist. Ogden and Richards speak, in order, of 'thought or reference', 'symbol' and 'referent', and they mean the symbol as the word and the referent as the thing (Ogden and Richards 1989: 11). Every, or almost every, semiotician uses different terms for these. But all agree that the relation between signifier and referent is indirect, while that existing between these two sides and the third side corresponding to the signified is instead direct. Thus, the base of the triangle is designated with a broken line, and the sides with a continuous line.

In other words, the semiotic triangle is arranged exactly according to the schema of cartographic triangulation, since this is similarly based on the opposition between the nature of the relation which links the ends of the base and that which belongs to the relation of each of these with the upper vertice. Simultaneously, there is, in this respect, an evident inversion which better than all else expresses the effectiveness and power of the cartographic sign over every other, and explains the nature of modernity. In triangulation, the base was crossed patiently, step by step, and, therefore, the relation between its ends was concrete and direct, just the opposite of what happens in the semiotic triangle. But this only means that in the five centuries elapsing between Alberti's first triangulation and the first semiotic triangle, the visual relationship, that is, the cartographic gaze, becomes the prototype for direct relation, to the detriment of the relation which involves the entire body. *Quid tum*, meaning: And now what happens? This was the motto Alberti had placed underneath his flying eye. The semiotic triangle provides the answer.

13. *The Logical Triangle*

Ogden and Richards' analysis aims, essentially, towards an impossible goal: to eliminate from natural language, the language which we habitually use, the ambiguity of the relationship between word and thing, to reduce this relation to the bijective relation that exists only in cartographic representation, in which all names are proper names, signifying, in a direct and unambiguous manner, the object they refer to. For this, more or less consciously, the authors take triangulation as their model; the process of the production of the modern cartographic image. It is triangulation indeed which establishes that every thing corresponds to a point, to a vertice, and that there is one, and only one, name for every vertice, exactly as in the triangle which we trace on a sheet of paper, there is one, and only one, corresponding letter for every vertex. The sole difference is that in the geometric triangle, this correspondence is absolutely arbitrary and conventional while in the cartographic triangle, concrete elements on the Earth's surface correspond to the vertices.

For Gottlob Frege, the founder of modern analytic philosophy, proper names (but also utterances) have not only a reference but also a sense (1892; It. trans.: 103, 106). Again, the cartographic triangle has the task of illustrating the difference between the former and the latter. Let the straight lines linked by the vertices of a triangle to the centre of the sides opposed to them be a, b, c, explains Frege. The point of intersection of a and b is the same as the point of intersection of b and c. The reference of the point in question is, therefore, single. Not its sense, however, meaning the manner in which this point is given, the mode in which it is presented. The sense is double, precisely because two possible designations exist: 'point of intersection of a and b' and 'point of intersection of b and c'. Precisely this uniqueness of reference and this plurality of sense guaranteed the effective cognitive content of a proposition for Frege. But here the point, we should add, is another: in illustrating the distinction between sense and meaning, Frege does no more than describe the first act of the cartographic operation which, at least from the seventeenth century onwards, is indicated by the expression 'filling in the triangles'. This consists in the insertion of the greatest number possible of minor triangles, thus of points, within the basic triangular grid, so as to transform the original skeleton system into a truly topographic image. This, again more or less consciously, may be just the model which is suggested by the analogy which the same Frege uses to clarify the relation between meaning, sense and mental image. Let's say that, Frege writes, an astronomer observes the Moon through a telescope. The Moon, the object of observation, is comparable to reference. The optical image projected by the lens on the inside of the telescope corresponds to sense: it is a partial image, it has one side only, and it depends, moreover, on the point of view of the observation, but it is objective, since it remains the same even if the observer changes. The image which in this case changes, and which is, therefore, subjective, is only the final image, the retinal image, which is individual and specific, that is, different, depending on the eye which looks.

It is sufficient to substitute the Earth for the Moon and binoculars for the telescope to transform the astronomer into a cartographer. So, we can recognize in Frege's analogy something of the thought of Anaximander, who, according to tradition, six to seven centuries before Christ, first dared to represent the inhabited Earth on a tablet (Farinelli 1998). For Anaximander, as for Frege, what we see arises at two removes from reality. For Anaximander, things and the things-which-are exist. We cannot fully know the first and, in the language of Frege, they are equivalent to reference. We can know only the second, the senses of the world, even if what we see is not the appearance of things (the senses precisely) but the appearance of the appearance of these, the appearance of the things-which-are. At least until the twentieth century, all Western philosophy remains blocked by the limit indicated by Anaximander, by the impossibility of knowing what Kant called 'things in themselves'. It is Frege himself who provides the explanation of the phenomenon:

precisely because senses are public and common that thoughts can pass from one generation to the next.

14. *The Logical-Semiotic Triangle*

Like Anaximander, Kant (§ 27) was also a geographer before being a philosopher. What is transmitted from Anaximander to Kant to Frege is the cartographic nature of the senses of the world, that is, the reduction of knowledge to the description of geographic representation, the map or the chart, however we choose to say it. What was for Frege equivalent to sense, the optical image of the telescope, is the ideal form of the map. Both have in common two-dimensionality, as well as the dependence upon a determinate point of view. But no map equals the precision of the figure reflected on the lens, even if it progressively approaches it through the continuous refinement of the instruments of surveying. Only in the second half of last century, with aerial photography and then satellite photography, does the map succeed, transforming itself into photography, in coming close to its model up to the point of becoming almost indistinguishable from it. In place of topographic maps, today orthophotomaps are spreading, recovered from aerial photographs onto which are added curves of level, that is, the imaginary lines which link up all points which have the same height with respect to sea level. Though before ceding place to photography, triangulation inspired, between the nineteenth and twentieth centuries, the most detailed schema which we have today of the relations internal to the representation of a physical object by a sign. This is thanks to Charles Peirce, perhaps the most acute intellect ever born in the United States, certainly a philosopher well trained, among other things, in cartographic technique (Brent 1993: 54). Peirce distinguishes three types of reference: the icon, the index, the symbol (Hartshorne and Weiss 1978: 156–73). Taken together, these form a hierarchy, in the sense that each corresponds to a level which presupposes the functioning of the previous.

Mental images correspond for Peirce to the iconic level, in which the relation between sign and object is based on resemblance. But resemblance does not produce iconicity. On the contrary, only after having recognized an iconic relation is it possible to hope to be able to distinguish what unites two things or groups of things. The iconic is the stage which corresponds, in the procedure of the cartographer, to the attitude of those who intend to fill in the triangles. This is the premise, not at all scientific, on the basis of which all other forms of cartographic representation are constructed. As such it is truly a presupposition, a preliminary interpretative option, as Ritter would have said (§ 1).

If at the iconic level, mediation between sign and object is assured through resemblance or similarity, at the level of the index it is assured by a relation of contiguity or correlation. It thus emerges from the assembly of several iconic relations, and from the connection—of a physical or temporal character—which results from it. When it is said that something is index of something else (the smoke of the fire, for example) it is meant that something is connected to something else from a causal point of view, or else that it is associated to it in space or time. This is the case, we might suggest, of the thermometer which indicates the temperature of the air, or the protractor which in triangulation measures the angles.

The third level of interpretation coincides, for Peirce, with the symbolic level. In this the relation between sign and object is assigned to a completely abstract convention with respect to the physical characteristics both of the object and the sign in question. Frege's distinction between reference and sense depended on the fact that words refer to objects (and then we have reference) or they refer to other words (and then we have sense). For Peirce, instead, the symbol is the product of a logical or categorical generalization which results from the recognition of an iconic relation between two different systems of indices. In other terms, the symbolic level is nothing other than the result of the systematization of a minor triangle in the pre-existent basic cartographic triangle, or else of the construction of another triangle adjacent to this basic triangle. It is, in short, the final product of cartographic work, made possible only by the acceptance of all its presuppositions. But where does this triangular obsession derive from, that is, this cartographic obsession, on which the whole of Western knowledge bases itself, and which will lead Peirce towards failure and despair?

15. *The Triangle and the Pyramid*

According to tradition, the first to measure the height of a pyramid was Thales, Anaximander's teacher. It is clear that if Thales had not already first reduced the pyramid to a triangle, the problem would not even have been posed. We might call to mind in this respect the fact that the pyramid has, properly speaking, no height. Height is, in fact, a dimension entirely foreign to it, in the sense that not only is it not seen but it is also impossible to calculate it directly. It is perfectly possible to climb a pyramid and to measure the distance from the top to the ground, but this distance corresponds not at all to its height but, rather, to the height of one of its sides. Otherwise, it would not be a pyramid but a tower. Thus, posing the problem of the height of the pyramid signifies presupposing a relation at the iconic level, exactly in the sense already recalled, between the pyramid and the triangle, the flat figure which most closely resembles it, and in which height is directly and immediately calculable. This relation involves the metamorphosis of a three-dimensional object into a two-dimensional schema, just as in every projection, the procedure which at least from Ptolemy onwards constitutes the original act of geography (§§ 4–5).

Diogenes Laertius reports that Thales measured the height of pyramids basing himself on their shadows, after having observed in which moment the shadow of the human body is the same as its height. Plutarch's version is a little different. In place of the human body there is a stick, the *gnomon* which, as Heron explained, 'makes everything similar' (Zellini 1999: 32–3). What does not vary is the shadow, the one brought to Earth by the pyramid and that brought to Earth by an object of which the measurements are already known. It is the length of two shadows, measured in the same instant, which allows one to obtain, indirectly, the height of the pyramid, in the form of the unknown of a proportion in which the other three values have been noted (the height of the pyramid is to its shadow as the height of the stick is to its own shadow).

In reality, things are, technically, a bit more complicated (Casati 2000: 98–101). But the fact that tradition passes on this story and not another means that its significance is not just technical. Thales passes into the history of geometry for having demonstrated the similarity of triangles whose angles are equal and sides are proportional. That is to say, without Thales, no triangulation would be possible. And his legend is valuable precisely as a description of the cartographic act as the original process of Western knowledge. Knowledge is born from the triangle because the carried shadow, which allows one to measure what cannot be seen and to assign value to the hidden, if not absent, dimension, is of triangular form: the first side is the body or the stick; the second, the shadow; the third, the ideal line which connects the tip of the first with the edge of the second. As the relation between shadow and surface of the Earth (the result of projection) is located on the iconic level, the proportion which is eventually arrived at, and which allows for the calculation, is situated entirely on the plane that Peirce called symbolic.

Heidegger wrote that normally the shadow is considered the absence of light, if not its negation, but in reality it is 'the mysterious testimony of hidden illumination', that which 'removed from representation, nevertheless becomes first of all in the entity, attesting thus being in hiding'. Have no fear: he is unconsciously describing Thales' measurement in his own language, in which the being (the pyramid) equals Anaximander's *thing* and the *entity* (the shadow) equals the *thing-which-is*. But why does what is true of the pyramid and its shadow also have to be true for the stick and its shadow? Why does the same relation which exists in the first case have to exist also for the second? The reply is based on the proximity between two sources of shadow, on their closeness. It, therefore, depends on a new conception of two places, that where the pyramid is and that where the *gnomon* is, in a single space (Serres 1993; It. trans.: 161–76). We have just seen, however, that without space there cannot be any relation of the index type. Thales thus illustrates, with his movement, the first indexical relation, in virtue of which all the others will function.

16. *Instructions for Use*

For Peirce, every thought was reducible to a diagram, and to think complexity equalled an infinite process of triangulation, not on one but on three levels. The words with which he describes the most complex level, capable of giving significance to events, all count, singularly or together, as synonyms of the cartographic image and its function: 'mediation', 'goal', 'generality', 'order', 'interpretation', 'representation', 'hypothesis' (Brent 1993: 331–2). However, reality today no longer obeys the logic of the map based on the syntax of the rectilinear and on the principle of the uniqueness of the centre. Mediation, goal, generality, order, interpretation, representation, hypothesis constituted by cartography (space in a word) no longer correspond to the way the world functions. In the case, for example, of the production of computers, the distance between locations where different phases occur holds an insignificant place in the determination of its price (Castells 1996; It. trans.: 446–53). What we today call globalization is nothing other than the totality of processes whose activity is unregulated, and thus uninterpretable, according to the categories of space and time which for the modern era governed the comprehension of what happens.

If, therefore, the equivalence between world and cartographic image of the world is lacking today, if the latter is no longer reducible to a map and, consequently, the cognitive process is no longer comparable to triangulation as we know it, we are brought back again to the foot of the pyramid. At its summit, we can then trace a 'W' which stands for world.

At the bottom, corresponding to the lower vertices of the triangle which represents the side turned in our direction, we can trace a 'P' on the right side which stands for place and an 'S' on the left side which stands for space. Further, still below and coinciding with the last and farthest vertice which it remains to name, that of the lateral triangle which has a side in common with the first triangle and is thus comprised of 'W' and 'S', we can trace a 'T', which stands for territory. It is important that the reader sketches this figure in their own hand because, as Edmund Husserl (1954; It. trans.: 65) already warned, every geometric construction implies a causal system, a hierarchy of causes and effects. Thus, to realize one's own drawing is the only way to not inadvertently suffer its effects, and to attempt instead to control them. For the same reason, this book has only two figures but no map, just as there were no maps (this is not a comparison but an appeal to authority) in the texts of critical geographers such as Alexander von Humboldt and Carl Ritter. For too long it was believed that geography was knowledge related to *where* things were, without realizing that in reality, in indicating this, geography decided *what* things were. It decided this as cartography did, that is, implicitly and silently, precisely appealing to the absolute power of the map which does not admit critics nor corrections. This does not detract from the fact that for the better comprehension of what will be said here, it will be useful to consult an atlas, and, before this and better still, a globe. But only in order to situate things, phenomena and processes in their correct position, whose nature we will already have tried to clarify, independently from their cartographic image.

We have already given a definition of the world, of place and of space (§§ 1, 3). The territory, which will be discussed in the volume dedicated to political and economic geography, equals a field distinguished by the exercise of power, and it is a word in whose root Earth (*Terra*) and terror are mixed and confused. Place refers to the iconic type of relation understood in its most general and elementary sense, and corresponds to the subject of the cognitive process. Space refers to the indexical type of relation, and corresponds to distance and its measure. Territory refers to the symbolic type of relation, and corresponds to the object. All this is the case since only a comprehensive theory of knowledge is able to substitute other models of geographic description for models of cartographic origin, today when the map no longer represents the world and we are thus constrained to do without any habitual interception against it. What follows is a first step towards the only possible global geography: the geography of the senses, of points of view, of world models.

PART TWO

The Landscape and the Icon

17. *The Name of the Mountain (and the Name of Ulysses)*

Still today the story is told, in Italian manuals of cartography, of the topographer sent to Bergamasco by the Military Geographical Institute of Florence, the institution to which we owe the cartography of the Italian state from its very beginnings. 'What is the name of that mountain?' the topographer asked a peasant working the country. It is easy to imagine that the topographer accompanied his request with a gesture, pointing his index finger towards the relief which he meant to carry the name on the map. 'So mìa,' responded the peasant, which in Bergamasco dialect means 'I don't know.' The good topographer, who did not know local dialect, wrote Mount Somìa, and so it is today read on maps (Bonapace 1990: 14).

'Names are petrified laughter,' wrote Max Horkheimer and Theodor W. Adorno (1947; It. trans.: 87), to signify the arbitrary character of all naming. On the map, where only proper names exist, each of these is the product of a systematic falsification, analogous to that just described. This occurs because every name on the map is the crystallization, the objectification—that is, the transformation into an object—of a relation, of a process that, as such, involves in reality the presence of at least two terms, two intentions, two cultures that do not understand each other and are often contrasting. Thus every name is indeed, just like every forced translation, invariably misleading. The cases could be multiplied at will. Even the word 'kangaroo' seems to derive from the response (absolutely identical to that of the Bergamasco peasant) of an Australian aboriginal questioned by an explorer on the name of such a strange animal. Topographers still cite the name Mount Somìa as an amusing exception, a rare error. In reality, it lays bare the normal mechanism of every cartographic denomination. Since without cartographic representation, the proper name does not exist (Bateson 1979; It. trans.: 47), this mechanism coincides with denomination in an absolute sense, *tout court*.

This is exemplified by the famous example, narrated in the *Odyssey*, of the name that Ulysses declares to Polyphemus. The success of the trick of the Greek hero depends upon its relation to a monster who is illiterate, illustrating the superiority of written culture (that of the topographer) over the simply oral culture which unites the giant to the Bergamasco peasant and the Australian aboriginal. Ulysses says to the Cyclops that he is called nobody, so that when Polyphemus invokes help because Nobody (proper name) has blinded him, the other giants do not understand and thus do not rush to him, meaning that no one (indefinite pronoun) is threatening their comrade. The equivocation (the oldest in Western culture) depends on the fact that in Homer's text, different to in Italian, the two words sound identical but are written differently: *Outis*, with the upper-case letter and as single word, and *ou tis*, with the lower-case letter and as two distinct words. It will be seen, following this, how Ulysses thoroughly represents the cartographic mentality. Let us note for now that he, exactly like every cartographer, assigns a name in totally arbitrary fashion, to himself. And that the trick of the invented name functions because he abolishes, just like the cartographer above, all separation between two distinct words, so that the interlocutor is deceived. This is exactly the same operation implemented with the blinding of Polyphemus, through the violent insertion into his eye of the olivewood stake from the incandescent point: the abolition of all distance, the suppression of every interval between here and there, by means of the indication with a gesture of a thing which cannot correspond to any verbal description. This is exactly like the case of the finger directed towards the mountain or the kangaroo, of which Ulysses's stake represents the cruel archetype.

Exactly this gesture (that of the pointing finger) was for Peirce the first example of the indexical relation (the index finger, indeed) between the sign and the object, based on a real physical force capable of capturing the attention of the gaze and leading it to stop in front of the object selected. But this gesture founds space (§§ 14–15), and it has nothing to do with place, if not in the sense of overcoming it.

18. *Landscape and the Icon*

The proof is, in fact, in the name of Somìa. The peasant who lived at its foot did not know its name because that mountain was part of his place, of the area within which he lived and worked and created his whole world. There not being for him another mountain, and thus not having necessity to distinguish between one mountain and another, Somìa was for him not a mountain but *the* mountain, the only possible mountain. As such, it did not have need of any further specification, such as a proper name. On the contrary, for the topographer whose only issue is to create a map and thus to distinguish between one mountain and another, the name which is determined exemplarily expresses the initial condition which is instead linked to his own ignorance, and says absolutely nothing about the mountain itself. Some years ago, Denis Cosgrove distinguished between the figure of the *insider* and the *outsider* (1984; It. trans.: 38, 246–7). The first corresponds to the inhabitant of a place, for whom what is seen corresponds to the area in which he or she lives, and which thus has no need of fixing on any index, if not in order to signal, if necessary, something he or she already knows. The second is the person who arrives from the outside, and, like the topographer, seeks to reduce what he or she sees for the first time to what he or she already knows. Actually, Cosgrove introduces this distinction with regard to the landscape, thus not precisely with regard to area but to a particular manner of perceiving, a specific modality of its representation. He notes that for the *insider*, strictly speaking, the landscape does not exist, since whoever lives in a place and knows nothing else cannot have consciousness of any diversity, thus not even within modes of considering the surface of the Earth or a part of it. As in the logic of the *insider*, there is no name for things, so too there would not exist, strictly speaking, the problem of the modality of conception of these same things.

The issue deserves proper consideration, all the more since it supposes a double ideal subject (the *insider* who is inside and only inside, and the *outsider* who comes from the outside) whose existence in a pure state is, especially in the first case and at the present day, somewhat problematic (§ 38). In the meantime, we can establish, however, what appears implicit in what we have already said. Different from place, landscape is not composed of things but is only a manner of seeing and representing (looking at) the things of the world. It is the form through which, in the modern era, the world comes to be seen from the point of view of place, as if the world were simply a local area or a collection, a series of local areas. The landscape is, then, the manner in which modernity conceives the world in the form of place, a representation which obeys a relation of the iconic type and excludes on principle both the indexical level and the symbolic level. It is, taking up Peirce's definition again, a mental image, of which the cartographer rids him- or herself as soon as he or she starts work, but without which no modern cartographic image would be able to be created.

In order that a landscape exist, three—not two—things are necessary: not only a subject that looks and something to look at but also the maximum horizon possible, thus a high ground which functions as a vantage point, or at least not to be in an absolutely flat place. The last condition is the most important. Although it may be surprising, only at the beginning of the nineteenth century, with Ritter's *Erdkunde*, did the geographic description of the world begin systematically to include the description of forms of relief which were almost entirely ignored until the end of the seventeenth century. Before the second half of the eighteenth century—before, that is, that pressurized barometers became reliable instruments—few people thought of measuring the height of a mountain (Dainville 1962). Thus the landscape presupposes not only modernity but also the domestication of the mountains, their inclusion in the *ecumene* which indeed only occurs in the eighteenth and nineteenth centuries. The subject of the landscape, the figure who looks from high down on the panorama below, is, therefore, a historically determinate subject. In geography, it coincides with the birth of 'civil society', of public opinion, which is opposed in Germany to the aristocratic-feudal world.

19. *Humboldt's Gift: The Concept of Landscape*

The concept of landscape enters to take part in geographic analysis thanks to Alexander von Humboldt, the other major representative, together with Ritter, of the *Erdkunde*. In the second volume of his principal work, the *Cosmos*, which appeared in Berlin in 1847, the year before the movements which will bring the bourgeoisie to power, he traces the story of the models that have governed, from the outset, the vision of the world on the part of humanity. The whole reconstruction revolves around the strategic value held in the model of the landscape. In this regard, Humboldt distinguishes three stages of knowledge, three steps in the cognitive relationship between man and his environment, valid not only in terms of phylogeny, that is, the history of the human race in its totality, but also in terms of ontogeny, the history of the single individual (1845: 4–6, 8–10, 17–20, 24, 66–7, 71).

The first stage is that of suggestion (*Eindruck*) which arises in the human soul as originary manifestation, as primordial feeling faced with the grandiosity and the beauty of nature. Its cognitive form is indeed that of the landscape, which corresponds to the world understood as harmonic totality of the aesthetic-sentimental type, which every rational analysis is (still) foreign to and which thus refers solely to the psychic faculty of the subject. *Eindruck* is a composite word, simple only in appearance. *Druck* properly means 'impression', and counts also for the impression of typographic characters on a sheet of white paper. For Humboldt, it instead involves the sensibility of the looking subject: the white sheet is its soul and the features of the landscape are the characters printed upon it. But the other half of the term is equally important, the prefix *Ein*. It signifies 'one', but has in reality a double function. On the one hand, it refers to singularity, to the individuality of the subject which looks and, in looking, initiates the process of knowledge. On the other, it signals the attitude of the subject to reduce the heap of impressions to unity, so that the cognitive field, from the very beginning, and even if only on the aesthetic plane and the plane of impression, is configured as a totality, as a whole arranged for the disclosure of the order 'hidden beneath the skin of phenomena'.

It will be the task of the successive stage, that of *Einsicht*, that is, of insight, to disarticulate the sentimental totality and initiate translation into scientific terms. In the word *Ein-sicht*, in fact, the prefix, which in appearance is identical, signifies the opposite of what it expresses in *Ein-druck*. *Sicht* here means 'view', the gaze strictly connected to reflective elaboration, to rational thought. The uniqueness expressed by the prefix regards not the subject but the object, referring to the concentration of thought on a single element among those present, in the form of totality, into the good first impression. In the intermediate stage, which is that of scientific analysis, there is no longer landscape (feeling, aesthetic impression) nor, consequently, totality, but only the cold and rational dissection of single components.

The eclipse of totality is, however, temporary, concerned only with the second of the levels of knowledge. Totality is re-established in the third and final stage, that which Humboldt identified with the concept of *Zusammenhang*, or, precisely, of totality constituted by the being together (*Zusammen*) in a relation of interdependence of all the elements previously analysed. This means synthesis, the point of arrival, the last term in the cognitive process. In this, thanks to the mediation of the analytic insight, originary totality is transformed and restored, no longer on the aesthetic plane and the plane of sentimental impression but on the scientific plane. The development of all knowledge, for Humboldt, is nothing other than the translation, finally, into scientific terms of a dawning impression expressed precisely by the landscape which is absolutely not scientific but without which all science would be impossible.

In the language of today's science, Humboldt's *Zusammenhang* corresponds to complexity, indeed, to global complexity. It is indubitably the case that when the history of global thought will be written, that is, the history of globality, Humboldt will enjoy a place of absolute importance.

20. *Landscape Is the Icon*

Between Humboldt's triple-cognitive schema and Peirce's triple schema on the relation between the sign and the object, there is an obvious analogy, since, if we look carefully, there is a precise correspondence between the stages of the first and the modes of the second. However, in the case of the first level, that of the landscape and the icon, we are talking of a true coincidence, of a programmatic identity. Humboldt's strategy is based on this very identity, directed towards the transformation of the man of taste into an observer of nature.

Its true goal was that of tearing the German bourgeoisie away from its 'vacuous poetic games', as Franz Mehring will later say in 1910 (1947[1910]; It. trans.: 164), to provide it instead a knowledge which is capable of guaranteeing, through scientific knowledge, control of the world. All this beginning exactly from the 'distinctive literary education' of the 'highest popular classes', of 'all the educated strata' (Humboldt 1845: 18). The last expression translates the German *die Gebildeten*, which literally means 'those who are formed by the image (*Bild*), by paintings'. The work will, indeed, be called *Aspects of Nature* (*Ansichten der Natur*), first published in Tubingen in 1808, which will convince the whole European bourgeoisie to study the physical world. It is with this work that the concept of landscape changes, for the first time in Europe, from aesthetic concept to scientific concept, passes from artistic and poetic literature to geography, and is charged with a new and literally revolutionary meaning, from the point of view of the history of knowledge. Even today the term *Ansicht* (literally, 'sight') means two things: what is seen and what is thought, signalling their unified sense, their originary inextricability, their absolute functional coincidence.

The field in which the realization of the Humboldtian project takes place is the field constituted by the totality of the bourgeois public sphere. The first to introduce the modern concept of public opinion in Germany, on the threshold of the last decade of the eighteenth century, was Georg Forster, with whom Humboldt travelled to Paris in 1790, to understand the republican revolution. Forster would return a convinced Jacobin, asserting the need for the immediate replication of the French experience in Germany: the failure of the Mainz insurrection, which he saw first hand, was only some months previous to his death. Humboldt decided instead not upon political but cultural revolution, hinging precisely on the concept of landscape and on the structural mutation of its function from aesthetic to scientific. This is a mutation that could only be realized starting from the artistic image, the sole image of nature then known to the bourgeoisie. It was necessary to lead the protagonist of the literary public sphere, the expert in the work of art, towards a vision of the world which could be developed into scientific understanding of the world itself, and which would no longer stop at the stage of simple contemplation. It was necessary, thus, to transform bourgeois culture starting from its aesthetic matrix, to change pictorial knowledge, to which that culture was limited, into natural science, suited to domination and not solely to mere representation. The landscape, the pictorial view, was, with Humboldt, the instrument of this transformation. For Humboldt, indeed, strategist of critical bourgeois thought, the entrance into the field of scientific knowledge presupposed a total traversal of the 'realm of aesthetic appearance'. The concept of landscape itself, which all the bourgeois recognize because they recognize paintings and other artistic descriptions, was conceived as the most appropriate vehicle to assure this passage (Farinelli 1991).

As a child, Georg Forster had accompanied James Cook on his second voyage around the world. When Humboldt would indicate the origin of his ardent desire to visit tropical countries, he would recall, along with the descriptions of the South Sea islands on the part of his friend, the paintings of William Hodges seen in London at the residence of the viceroy of India, Warren Hastings, which depict the banks of the Ganges. It was still a period in which, to use the words of Heinrich von Kleist (1810: 76), explorers circumnavigated the globe 'to see if it was not perhaps more open at some point from the back'.

21. *Humboldt's Gaze and the Cunning of the Picturesque*

Humboldt never managed to go to India. For the English, it was unthinkable that a German spy, insofar as great friend of the Prussian king, wander undisturbed in their domains. So, in the end, Humboldt turned to the tropical world of America, employing all of his substantial maternal inheritance in the last great voyage without scientific exploration, after which the organization of the reconnaissance of the globe became, at such levels, a state affair. Between 1799 and 1804, he advanced to the Orinoco basin, and to the Andes, climbing towards the American isthmus from Cuba to Washington which Thomas Jefferson then finished planning and building. As Chateaubriand, the first French romantic, argued, in America, Humboldt 'painted everything and wrote everything' (1850[1811]: 60). More recently, Antonello Gerbi explained that 'with Humboldt, Western thought finally achieves peaceful conquest and ideally annexes to its world, to its single Cosmos, those regions which were until then almost only an object of curiosity, wonder or derision' (1955: 453). A true revolution of the gaze, thus, which collects and makes use of the lessons of the 'picturesque travellers' who are especially attracted, at the beginning of the French Revolution, by Mediterranean volcanoes whose 'bewildering wonder' not even Goethe (1903; It. trans.: 168, 200) manages to escape. Already for them, the world consisted literally of a series of paintings whose description depended on the prior reduction of the features of the Earth to a group of artistic illustrations. So, for example, Jean-Pierre Houël describes the departure from the Campania coast towards Sicily: 'Every instant new objects were offered to our sight. Close to Vesuvius on our right, Pozzuoli and Baia drew near on our left after the departure of our ship from the bay, and formed a single picture with the city of Naples' (1782: 2).

This, precisely, is the picturesque image, as is explained in the respective entry in Denis Diderot and Jean le Rond d'Alembert's *Encyclopédie* (1751–66). This is an image in which the glance has a great effect, 'after the intention of the painter', although, at the same time, objects are easily distinguished, at the cost of reducing the dimensions of human presence to the minimum. The function of the latter is only to make appreciable, by contrast, the immeasurable dimensions of the natural scene which makes up the background. Thus a pair of small figures allows one to leave almost all the foreground to the minute and precise rendering of exotic flora and fauna, to the unusual morphologies (for the European reader) whose representation is the true goal of the artist. In the magnificent folios of the travels in the Kingdom of Naples and islands by Houël and Jean-Claude Richard de Saint-Non, picturesque images consistently complement text in an organic relationship which makes them inseparable. The monumental report of Humboldt's American trip, in 35 volumes, was accompanied by two atlases, one geographic and the other composed of views: the *Vues des Cordillères et monuments des peuples indigènes de l'Amerique*, published in Paris in 1810. These colour engravings, in which artistic canon and scientific illustration become one, were the most subtle and incisive instrument of Humboldtian strategy, since the landscape, for Humboldt, coincides accurately with both of these.

We open the book, for example, at the central panel, which depicts the Chimborazo seen from the Tapia Plateau. Indian figs, cacti, rocks and llamas, together with indigenous people in colourful dress, stand out on the plane dominated by the volcano which was then believed, with its height of nearly 6,500 metres, to be the highest mountain in the world. Its mass looms mightily against the cobalt blue of the sky, made even more vivid by the white cap that adorns its peak and descends along its sides until disappearing, with an absolutely straight edge, below a certain altitude—the permanent snow line, which Humboldt was the first to investigate systematically, and which most of all presses upon the viewer to observe, enraptured by the beauty of the scene and the brilliance of the colours. (What is true of these drawings is also true for the geometric schemes mentioned earlier: no reproduction can substitute direct vision, not so as to avoid giving in to its implicit content in this case but, on the contrary, to be able to truly remain seduced by it, precisely as was Humboldt's intention.)

22. A 'Nebulous Distance'

Goethe himself recognizes, in *Elective Affinities* (1809), Humboldt's extraordinary ability and his capacity for seduction. The story of their relationship is not quite so simple and is bound up with, apart from silences, differences and agreements. For example, Goethe, contrary to Humboldt, will never allow the use of instruments which serve to help in the investigation of nature, and will maintain that the eye is sufficient as it is for this purpose, without resorting to any prosthesis. But on one point both agree: on the presence, in every view of the landscape—that is, whenever one looks at the world from the point of view of the scenic—of a certain haze on the horizon, a progressive loss of clarity and clearness of the air as the distance increases. This is not a simple atmospheric phenomenon connected to particular climatic and meteorological conditions, as one might at first be led to believe, but a cultural and political fact. For the Goethe of *The Italian Journey* (1816–17), the horizon appears hazy because his gaze is strongly influenced by paintings by the landscape painters he admired in Germany before his departure, among which there are some whom Humboldt had taken as models in the *Cosmos*: Hannibal Carracci, Saloman van Ruysdael, Allaert van Everdingen, Nicolas Poussin, Claude Lorrain (Hard 1969). In the case of Humboldt, the question is still more complicated, because what in Goethe is not completely conscious and in any case unintentional, is instead incorporated into a precise and deliberate project, following the course of a calculated and coherent metaphor.

For Humboldt, as for his bourgeois fellow citizens, the fascination of tropical countries depended, before all else, on the fact that despotic aristocratic-feudal power was totally absent in these countries but which ruled at home. Because this power, strongest in the lowlands, is weakened and even disappears in the highlands, the German mountain is for Humboldt, as for his friend Schiller, the home of freedom, a kind of domestic version of the tropics. It is from the peaks of the mountains that the 'nebulous distance' is manifested, provoking 'an enchantment full of mystery', an impression which is reflected, severe and foreboding, onto the spirit and feeling: the image of the 'sensuous-infinite' (Humboldt 1845: 38), of the fatally incomplete character of what we see, the structurally unfulfilled character of what we know, the programmatically partisan character (even if aimed towards totality) of what we do. For Humboldt too, as for Goethe, the haze in the distance enveloping things is the indicator of the dependence of literary description on pictorial representation; at the same time, it is much more than this. For Humboldt, who could even be called a political strategist of knowledge, it is the metaphor for all projective intention, for all political-social planning: always on the horizon and never reached, and thus indeterminate in its further forms. Just as Humboldt puts to use the ambivalence of the term 'landscape', whose signifier and signified are indistinguishable, so too the object and the concept, in the first stage of his cognitive strategy, present the same vaporous and indefinite form (Farinelli 1991).

As form and stage of knowledge, as phase of a strategy, as vehicle of a tension, landscape becomes, abruptly and without explanation, a simple group of objects, with the 'founding' of Siegfried Passarge's *Landschaftskunde* (translating as 'landscape geography') in 1919. With the publication of this text, landscape becomes equivalent to geographic reality, since the landscape is the sole form of reality accessible to geography. Mood and cognitive stage, made therefore invisible, along with what is *subsistent* (what cannot be touched nor counted), become in this way the thing, visible and existing, within the reach of sight and the touch of the geographer. The First World War was the opportunity and impetus for this new and sudden ontological change. The instrument was the camera, the tool which instantly reduces to fact and objectively reduces to product what was previously instead the result of a subjectively based and consciously determined process from the social point of view. In this way, consciousness of the processual and social nature of knowledge (not solely geographic knowledge) experience the same fate as the haze which represented them: they disappear to sight and thus cease to exist.

23. *Mountain and Plain*

Humboldt dedicates the *Aspects of Nature* (1849) to all 'oppressed souls', a definition which includes both those who seek the mountains as the kingdom of freedom and those who already live there. The apparent contradiction is explained by the fact that in the history of humanity, the mountain has often functioned as a place of refuge for minority cultures, for peoples thrown out of the dominant organization of settlements which has always been interested in the plain. This is already long before the birth of modern nation-states which typically crystallized an already defined structure.

We might think of the growing popularity, between the thirteenth and fourteenth centuries, of the heretical movements of the Cathars and Albigensians along the line made up of the Alps in Provence and the

Pyrenees. Or else, on the coast opposite to the Mediterranean, the Berbers, nestled among the mountains of North Africa (the Atlas Mountains) or the Sahara (the Hoggar Mountains), able to hold off all the invasions coming from the sea (Phoenicians, Romans, Vandals) but retreating from the front at the first and only raid coming from the land—the Arab invasion which ended at the beginning of the millennium. Or else, we might consider the cultural geography of the British Isles: inside these the Celtic element, which before the arrival of Caesar spread over the whole country, is again today confined to the highlands, carefully avoided not only by the Southern conquerors but also by their successors, the Angles and the Saxons arriving from Scandinavia five centuries later.

In fact, the opposition between plain and mountain belongs to the origin of Western culture, and the supremacy of one over the other is connected to the rise of the first cities. For the Romans, the distinction between *ager* and *saltus* was fundamental, between the flat and ordered field of permanent residences and cultivation (that is, something *of culture*), and the precipitous and disorderly mass of highland, realm of unstable sheep-farming and the absence of civilized values (that is, literally, those values related to the city). For the ancient Greeks, a single word, *oros*, meant both mountain as well as limit, a sign of the same sharp contrast marking all European feeling, though born in the Mediterranean. This contrast did not pass on without a curious effect of amplification and, almost, inversion: already between the eighteenth and the nineteenth centuries, travellers coming from the great European continental plain were awed, arriving in the Kingdom of Naples, to see towns and villages perched on the slopes, or on the Karst plateaus of the Apennines where settlements in fact thrive up to around 1,500 metres (Farinelli 2000: 130–1).

Less than 8 per cent of the entire world population, it should be pointed out, today lives above 1,000 metres while more than half lives below 200 metres (Ortolani 1992: 37). It is rare to find, at least in the modern period, an example of the submission of the plains to the reason or the economic needs of the mountains. The most extensive and clear case concerns the practice of transhumance, the seasonal migration of animals and men from the mountain to the plain and vice versa, from summer pastures up high to the winter pastures below, spread through-out the *ecumene* (Rafiullah 1966). Not even its most complex and rigidly regulated form—the form planned in the Iberian peninsula and in the Italian peninsula by Alfonso of Aragon in the middle of the fifteenth century—survived the subsequent growing demand for cereals, culmi-nating with the formation of national markets for agrarian products. In the course of the nineteenth century, the Tavoliere delle Puglie, until then a single gigantic winter pasture for sheep (and also for cattle and horses), was used to farm; from the consequent crisis, transhumant herding in Southern Italy would never recover.

When Humboldt, then, invites the oppressed to enjoy the freedom of the mountains, in reality he begins what some have defined as the 'sheep clearances' of the old continent, that is, the systematic submission of the European mountain economy to that of the plains. This meant the end of that same freedom of the mountains. Until this point, between the eighteenth and nineteenth centuries, highlands attracted attention and gained importance in the organization of state territories, in political and economic life. The reason for this was simple. They obstructed the very purpose for which nation-states arose and were consolidated: progress in the speed of the circulation of goods.

24. *If on a Summer's Night a Traveller*

This, however, does not detract from the validity of the cognitive func-tion of the concept of landscape in the times of Humboldt, even outside of the Humboldtian strategy. Let us, then, provide the necessary condi-tions for the existence of a landscape. These are: an observer, a relief, a panoramic view and a bright, if not perfectly serene, day. Below, in the valley, or on the plains, things can be seen one next to another, and close-ness allows one to assume with a certain assurance the existence of functional relations between them. The city, for example, exploits the coal field in its factories, which are only a small distance away, and if between the two there is a certain distance, the railway track and the white trail of steam emitted from the train's chimney clearly and immediately signal the existence of the link. In other terms: between the functioning of the world and what is visible, there is a general correspondence. If the same traveller had returned regularly, let's say every year, to admire the panorama, he would have been able to appreciate, perhaps with the help of a good pair of binoculars, the continual growth of certain elements, and thus the selective additions of some functions at the expense of others: the expansion of buildings instead of fields and meadows, the doubling in size of the railway line, etc. It must be noted that 'continual growth' here means a material development which not only knows no ceasing but which is also manifested through three already-mentioned (§ 4) attributes of Euclidean extension: continuity, indeed, but also homogeneity and isotropy (horse-pulled trams, for example, run between the centre and the edge of the city).

However, this was equally true in the times of Humboldt, in the period, that is, of the first industrial revolution, based on the factory system and the connected increase in urban population, on coal as an energy source and on heavy alloys (iron, cast iron) as a raw material. Let us now, instead, look at the example of a traveller who observes the same scene a century later, at the beginning of the twentieth century, in the period of the advent of the second industrial revolution, that based on light alloys such as aluminium, on chemistry and, later, on plastic and electricity. The posts and wires of light again recover a trace of the connection between energy source and place of its use, but this trace is a much smaller trace than in the railway line. The distance between the place of production and the place of use of the electricity supply can be very great, even in the order of thousands of kilometres. The greater or lesser distance between the things of the world no longer refers, or almost not at all, to the relations that govern their activity, since their closeness or distance almost no longer means anything, and is no longer the indication of any necessary relationship. Any plausible systematic correspondence, any immediate congruity, between the functioning of the world and what is visible to the observer's gaze is lost.

Thus a third traveller who reaches the same place and watches from above in our day would almost not be able to pick out any indication, any material trace, allowing one to judge the interdependence between the things which he or she sees. This impossibility is the product of miniaturization, dematerialization and informatization, that is, the joint application of telematics, cybernetics and electronics to production and communication networks. This results in a world in which, for the first time, the domain of vision no longer provides almost anything signifi-cant to the mechanisms which regulate the reproduction of the world's activity (§ 32). This is an enormous problem for Western culture which for centuries founded knowledge on vision, and which in the modern era made vision coincide with the certainty of representation. However, already before this, it was a crucial question of geography. Much more than when, at the beginning of the 1960s, Waldo Tobler (1963) made the last attempt, chronologically, to develop the laws of geography to demonstrate its character as scientific knowledge and succeeded in for-mulating only one of them. He recited this rule exactly like so: all objects which exist on the Earth's surface interact, but their interaction is stronger when they are closer and weaker when they are more distant.

25. *Herodotus in Berlin and the Mysterious Subject*

Tobler's law experienced the same fate as the Berlin Wall, constructed in the same period to separate the countries of Western from Eastern Europe: neither the Wall nor the law functioned for a single day, at least for the purpose in view of which they were made. In the case of the first, this was to obstruct the passage of the most precious goods from one side to the other: money and information. This happened for one single reason: informatization began in the very same months. Bringing these events together, from the beginning of the 1970s, the first electronic calculators began to dematerialize the world, transforming atoms into bits, thus putting the importance of material distance into crisis (§ 92). The model for Tobler's law dates back to Herodotus, although Tobler did not know this. In order to explain, in Pericles' time, the difference between Greeks and barbarians—those who did not speak Greek—Herodotus in fact made the degree of difference depend on distance. All those who are not Greek are barbarians, Herodotus explained, but the farther a people is from Greece, the more it is barbarous.

The Berlin Wall and Tobler's law were Herodotean because they took on the same logic as the author of the *Histories*, for whom the mechanism of the world depended on linear metrical distance and was, essentially, a spatial function. However, since, different from Herodotus' *Histories* (and the Wall), Tobler's law presupposed the existence of space and not of the world: it concerned only things and said nothing about men and women, about the inhabitants of the world itself. This silence decreed for geography the liquidation of all capacity for reflection on its own foundations, beginning at the end of the Great War from the *Landschaftskunde* of Passarge (§ 22) and continuing with particular intensity in Anglo-Saxon geography in the third quarter of the twentieth century. Landscape, from being an (exceedingly interested) way of conceiving reality, a strategic sense of the world in relation to change (as it was for Humboldt), became a collection of objects with Passarge, a series of elements. From a means of interpretation, it became, in this way, a simple group of features given objective form once and for all, no longer essentially dependent, in its constitution, on the ideational activity of a subject possessing a psychology and an intention, or a project. It is the existence of the subject, of a subject of geographic knowledge, which is negated in this movement, and in the most resolute form: it is abolished abruptly and in one fell swoop.

The history of the subject of geographic knowledge is a truly curious history, yet to be written. In the times of Humboldt and Ritter, its figure corresponds to the figure which, in the paintings of Caspar David Friedrich, the pioneer of German romantic painting, is always silhouetted from behind at the very centre of vast and desolate landscapes with boundless horizons. It was the same Friedrich who would explain this singular and mysterious pose in political terms, as a sign of his (own) belonging among the ranks of those that, at the beginning of the nineteenth century, had espoused the ideal of a united Germany, equipped with a constitution and governed by the agreement of its citizens. Beyond this political meaning, all the paintings of Friedrich transmit at least two further warnings about the nature of the subject, the landscape and their relation. These are: (a) that the subject is part of the landscape, an observer internal to the observed system and not external to it; (b) precisely for this reason, our vision is very limited, in the sense that we cannot see everything. For example, the face itself of the figure looking at the landscape in the painting. To see means to be bound by one, and only one, point of view, while in reality we are observed from all sides. In this way, the subject painted by Friedrich is much more complex than it appears: it represents the artist-traveller who stops during the journey and records through his presence that what we see is not nature but the experience of nature as we successively re-imagine it (Koerner 1990: 179–244). The landscape, the 'painting of nature' as Humboldt described, is the image that, through what is seen, does not just show what exists but announces what will happen just afterwards. This constitutes a real crisis of the existing political and social order.

26. The System of the Landscape

Alfred Hettner (1923: 49), the last heir of the critical-epistemological tradition of the *Erdkunde*, was the first to seek the definition of landscape in the work of Passarge. No such definition exists, because, at the beginning of the twentieth century, the ontological question ceases to exist for geography and, consequently, its epistemological response as well. The Earth's features, which until then were considered merely empty appearances waiting for a meaning which it was the geographer's task to assign, become, in this way, the only truly existing things, complimenting what Leszek Kolakowski (1966; It. trans.: 4–5) defined as the first rule of positivism. This is the rule of 'phenomenalism', according to which—different from what the *Erdkunde* still believed—there is no real difference between the essence and appearance of things. In this way, geography is reduced to the simple inventory of the forms through whose 'thinking consideration' the *Erdkunde* began the 'completion and criticism' (Kramer 1875: 375) of existing geographic knowledge. The instrument of landscape ends in being identified with the same object which comprehension originally had to serve. What was once the cognitive process is transformed into the thing to be known—it is thingified. To return to an image: the flight of the arrow disappears and there only remains the target, so that no relation is any longer possible, and no one any longer remembers the archer, even less his original intention. In other terms: the subject disappears immediately from the landscape— reduced to a group of things, to a complex of objects, no longer conceived as a dialectical model of knowledge or as a sense of the world —and, with it, all possibility for explanation disappears from geography.

To convince ourselves of this, let's open the book that still today remains, at half a century's distance and on the international level, the fundamental work of reference in the geography of landscape, understood in an analytic and systematic sense: *The Earth's Landscape* by Renato Biasutti (1947). It distinguishes, on the first page, between 'landscape able to be sensed or seen', which the eye can embrace in one turn of the horizon, and 'geographic landscape', the abstract synthesis of visible landscapes, composed of several recurring features, meaning more often 'above a quite large space, higher, in any case, than the space comprised of a single horizon'. In this way landscape is composed, in the eyes of the geographer, of a small number of characteristic elements or a few groups of elements, making possible not only its briefer description but also the identification and comparison of its main forms. Biasutti's analysis is based on the intertwinement of events related to four orders of phenomena: climactic, morphological (both as related to the structure of the Earth's surface, that is, endogenous, as well as due to modelling by exogenous agents such as courses of water, glaciers, the sea, the wind), hydrographic and vegetative. From the combination of these there emerge 34 fundamental types of landscape, which together exhaust all present forms on the Earth. Even today this synthesis remains unsurpassed—but at what price?

First, it refers exclusively to certain, not to all, natural features, and does not include cultural features in any way, even if the author, whose training was anthropological, dedicates the final chapter of the text to the illustration of the 'human landscape' and its realms. Human history exists alongside nature only in this sense, repeats Biasutti, basing himself on Ritter. At the same time, he continues, man is able to break free from the direct ties which bind physical and biological facts. This is a sufficient basis for the separation between nature and culture and, therefore, the specificity of the natural landscape relative to the cultural landscape and vice versa, even if very strong correlations exist. The greater cost, however, concerns the renunciation, seemingly temporary, of the 'search for links, causal or otherwise, between associated phenomena' (Biasutti 1962[1947]: 11, 8). This renunciation or, rather, suspension lends itself to two interpretations. The first concerns the danger of superficiality typical of any description that, like geographic description, is programmatically and literally limited to the surface of things. But this interpretation is, perhaps, in turn, excessively superficial. Another, more subtle, interpretation is possible.

27. A Geographer called Kant

Roland Barthes spoke of 'absence of causality' (1954: 33) with respect to the historiographic conception of Jules Michelet, who, in 1833, in the aftermath of the publication of the first volume of the *Histoire de France*, had exclaimed with pride that before himself no one had spoken of geography as a historian. This was not true, but that is not the point. Michelet, Barthes explains, did not know how to create anything apart from mechanistic explanations of historical processes, since for him the objects of history were 'different moments of the same section', or even 'two more or less distinct zones in the same stretch of water'. As such, they were irreducible to an order 'based on the opposition of heterogeneous objects like cause and effect' (ibid.). The same is true, in some measure, for Biasutti's landscape, since Biasutti was perfectly conscious 'that all phenomena and objects united in a given space on the Earth are reciprocally linked by some relation', by the fact, that is, that proximity contains an explicative principle (Biasutti 1962[1947]: 7). Only in Kant's *Physical Geography* was the same point affirmed (Kant 1807).

Kant wrote of philosophy but taught geography. He indicates the decisive turn in his thought in the passage from the empirical geography of the seen to 'the geography of reason', as he calls the critique of pure reason, that is, to the geography of 'the blank space of our intellect' (Cassirer 1918; It. trans.: 173–4). This is a turn in which the obscurity, that is, the invisibility of our mind, part of chthonic nature, is added to the immaterial character of the object under investigation. Even after this turn, however, the first thing that Kant taught to his students was the distinction between logical classification and physical classification. The first, that is, Linnaeus' taxonomy, which had put the plant world into an order, grouped species and genera of plants according to the resemblance of parts which tend to remain invariable in the course of evolution. In this, Kant recognized the benefit of having founded the economy of nature, but underlined its character as an inventory of 'isolated things', removed from their original context and combined artificially according to a logic. Plants, for example, that were far apart, typical of different regions of the world that were very distant, from entirely different environments, were found to be closely related as part of the same family only because their organs of reproduction exhibited some formal similarity. On the contrary, history and geography—which, more than sciences, are knowledges for Kant—proceed precisely on the basis of physical classification which follows the laws and order of nature. Geography too, in particular, 'represents natural things according to their species and their family' but, different from sciences which are part of the economy of nature, it represents them 'according to the place of their birth, or the places in which nature has situated them'. It represents them, that is, according to the principle of closeness or proximity, one thing next to another just as in reality they actually exist, just as Mediterranean plant life is offered as a single organic unity to our gaze. While according to Linnaeus' method, which still prevails, this is composed of essences that are not only distinct but that also belong to different species, genera, families, orders and classes (Kant 1932; It. trans.: *xi–xxiv*).

Kant does not proceed; he stops at the statement of the opposition between logical and physical classification, between the principle of resemblance and the principle of closeness, and turns to narrating the adventures of Cola Pesce who was able to stay hours and hours underneath the water in the Straits of Messina. The problem which he poses, however, is enormous and decisive—it concerns, in the final analysis, the reason for the difference between the scientific image of the world and what we instead have of it when we throw open the windows in the morning, when, in short, we consider the world as if it were a landscape. Only in this sense, in fact, only in the form of the landscape, are the things of the world given one next to the other, coexisting in their organic unity and perceived together, before any disarticulation or reflection. But as was seen with Humboldt (§ 19), this form is not at all scientific and is instead seen merely as prior to the scientific consideration of the world. Perhaps, thus, Biasutti too, who knew Humboldt as well as Kant, excludes all causal relations from landscape.

28. *Geographic Type and Ideal Type*

At the distance of a century, nothing remains of the Humboldtian project, not even the memory of it. At least in geography. The expulsion from analysis of the cause–effect relation coincides with the refusal of the harmonic character of landscape, of the aesthetic form that, according to Humboldt's strategy, was the only one capable of leading the bourgeois subject in its early stages towards scientific discourse. Nature, for Biasutti, is not harmonic or discordant: it is everything. All possibilities are open to it. A glacier in the forest, two enormous perennial 'tumboa' leaves in the most dreary of Australian deserts: Are these not perhaps discordant elements? In reality, he continues, harmony depends on the possibility of tracing the causes which have produced different phenomena, the rules which govern the game of their relations. Therefore, harmony is everywhere and nowhere, and, in any case, is connected as it is to the principle of causality, not solely related to landscape (Biasutti 1962[1947]: 8). All consciousness therefore disappears of precedence and of the activity constitutive of the (pictorial) model relative to the landscape itself, of the subject relative to the object and of intentionality relative to geographic knowledge. Geography also loses, if not all consciousness of the necessity of an orientating model for research, then any possibility of ideal-typical concepts in the sense given to this expression by Max Weber (1951; It. trans.: 107–25) in relation to the historical-social sciences. Even if in appearance one might say precisely the opposite.

As in Weber's 'ideal type', Biasutti's types of landscape, in their conceptual purity, can never be traced in empirical reality. These are also syntheses that, in order not to be contradictory, are achieved by the unilateral accentuation of one or more features, and the grouping together of others which are hidden if not actually absent. In tropical savannahs, for example, the mass of vegetation is composed of perennial graminacees, around a metre high on average, among which the *Andropogon* genera stands out, which along the Upper Nile may reach even 6 metres and can be considered the highest form of grass known. As such this too, even if quite rare, enters into the determination of the related type of landscape. The function of the latter is, in general, a schema with which reality is compared and measured with the purpose of illustrating specific significant elements of its empirical content. This, however, is precisely the problem. The typical-ideal character works at best for single landscapes but not for the concept of landscape. Precisely insofar as the former derive from the latter, the concept of landscape is returned to a validity which is not only logical but also practical, as Weber would say, in the sense that it functions as a model of what landscape must be according to the conviction of the author, that is, according to a value assumed to be permanent. It implies, hence, an evaluative judgement whose existence is totally independent from the greater or lesser level of awareness in which it is expressed. In the case of Biasutti, this judgement is hidden and is already, at the outset, contained in the idea of visible landscape in a completely unexpressed fashion. This is even before it is contained in the abstraction that leads to geographic landscape. Judgement consists precisely in the seeming independence of geographic landscape from every prior and more general option. Judgement consists, that is, in the implicit, seeming absence of every pre-judgement.

For Leibniz, it was impossible to perceive the world without at the same time judging it (1961; It. trans.: 430). There is no need to disturb the philosophers, however; the geographers will do. In 1921, Albrecht Penck wrote to William Morris Davis, follower of Ritter and founder of the theory of the cycle of erosion in geomorphology: 'Do we truly only say what we see? I think it is foolish to imagine that an observation can exist which does not involve an inference. A Kodak can observe' (Chorley et al. 1973: 523). That is to say, every gaze involves an evaluation; it is the result of a deduction and derives from a presupposition. For Biasutti, on the contrary, geography is an inductive science, proceeding only from the particular to the general. As such, it remains descriptive, and the geographic matrix remains, different to Weber's 'ideal type', absolutely unconscious and unreflective. It is not for this reason, we will see, less active and powerful. On the contrary.

29. *The Forms of Landscape*

Landscape—reduced from a way of seeing to a group of elements, from a vehicle of a social project to a series of material features, from a sense of the world to a collection of things, from a subjective projection to an objective complex of forms—immediately exhibits its own limits. These are illustrated at the beginning of the 1970s by Lucio Gambi (1973a), prior to and more skilfully than others, in an essay that, incidentally, represents the first of the very rare cases in which Italian geographic reflection after the war precedes international geographic reflection. Gambi's thesis is simple, aside from being well established: what does not have visible form moulds and builds that which is instead visible, so that the latter, which corresponds to landscape, is only a consequence of the former. Thus, we find that the concept of landscape seems absolutely insufficient to indicate reality.

This exemplification takes the up discourse exactly where it had been left off, at the beginning of the 1930s, by Marc Bloch, in the manifesto of modern agrarian historiography. Bloch had distinguished, in relation to France, between two agrarian systems. The first was organized around open fields (*openfield*), without enclosure and more or less extended in strips, or otherwise more squat and irregular. The second was based on closed fields, surrounded by low walls, fences or trees and hedges (*bocage*). The formal opposition, immediate at a glance and thus evidently scenic, was accompanied by the genetic opposition, and both were returned to a sharp and compact usage, able to be synthesized in schematic form. Open fields, distributed over all of Central Europe, dominate France east of the Havre meridian and north of the latitude of Dijon, and derive from practices based on collective crop rotation which was customary in the Middle Ages and which complimented the needs and decisions of the community more than the individual cultivator. The form of buildings [*incasato*], constituted by houses grouped together and never spread out on the fields, testified to the prevalence of the logic of the collective over that of the individual. On the Atlantic side of

Europe, and in particular in Western France, field enclosure prevails instead, signalling the absence of communitarian spirit in agricultural practice and in livestock breeding. The style of settlements, organized around isolated habitations or those collected into incredibly scarce clumps, duly confirms the traditional absence of collective work and communal solidarity, and identifies instead the individualistic imprint of activity (Bloch 1952; It. trans.: 42–76). If we wish to recognize an example of this contrast in Italy today, we might turn to the opposition between argillaceous maritime Abruzzo and calcareous mountain Abruzzo. Here, there remain open fields, centralized groups of houses, cereal cultivation and the survival of communitarian practices of late medieval origin, such as easements of way and easements for purposes of irrigation (the right, that is, of every peasant to pass or let his own water pass through other people's fields). The hills which face onto the Adriatic Sea have, instead, been the realm of closed fields for centuries, with houses spread out on smallholdings, and work organized in an individual manner through the intensive mixed polyculture characterizing Mediterranean countries.

Precisely through the insertion of the landscape of mixed cultivation, Gambi complicates the binary schema of Bloch, rendering it more articulate and applicable to regions south of the great continental European plains. An expression of very ancient skills, Mediterranean polyculture unites not just two (as occurs further north) but three levels of cultivation on the same field, one on top of the other. These are: the herbaceous layer, the shrub layer (the vine) and the tree layer. Their association gives rise to more sophisticated rural architectures, already described by the Romans at the beginning of the Christian era. Even if on the path to disappearance, the vine married to the olive and the ash tree (which is aligned to the vine and sustained by it in rows on the field) is even today, from Emilia Romagna to Lazio, the one sung of by Virgil. Different to the previous types of landscape, what is decisive for identification in this case is not the absence or the presence of enclosure, or the form of fields or houses, but the plot of the rows, of the 'piantate', whose form is the expression of a whole agrarian structure (Desplanques 1959: 29–64). It is, thus, the expression of a very original history.

30. *Anomalous Landscapes*

It is so original that to historians it seems, even today, absolutely anomalous. On closer inspection, the architecture of the 'piantata' or 'alberata' was (and still is today, insofar as it survives) a sort of expedient for the triplication of the agricultural land. It results in the almost paradoxical venture of combining three plants which are in fact antagonistic to one another. If to this we add that livestock grazed in the fields after the harvest, we must ask ourselves how mixed cultivation was regarded as extensive for such a long time, despite the sophisticated and intensive procedures it manages to use on the terrain. The response consists in the continental, and not Mediterranean, origin of historiographical and geographical models.

Some years ago, Maurice Aymard (1978: 1169–82) defined the half-millennium in Italian history which runs between the fourth century and the end of the nineteenth century as 'unclassifiable'—as compared to the classical English model of the transition from feudalism to capitalism. In the latter, the passage occurs between the seventeenth and eighteenth centuries and coincides with a pronounced and general industrialization. Therefore, the transition is clear and the succession is immediate. In Italy, instead, the communes of the centre-north see forms of capitalist development until the medieval period, forms which provoke, that is, a precocious process of 'feudalization'. This is not accompanied by, as in the British Isles, the development of large industry, except after the political unification of the peninsula. In the intervening period, which actually lasts for five centuries, the longest economic 'phase of indecision' ever known by a Western country is experienced, according to Aymard. These are precisely the centuries of the spread of

mixed cultivation in rows, which never surpassed the medieval boundary between the urban Italy of the Po Valley and central regions and the
feudal Italy of the south, except Naples. This expresses the indecision of
modern Italian agriculture, its way of compromise, its median and
balanced solution between production for subsistence (preceding all
capitalist development) and production for the market, typical of mature
capitalist systems. This Italian third way is responsible for the main
fascination and the attractiveness of the Italian landscape in the eyes of
foreign visitors.

The anomaly of the landscape of the piantata was based on the system of sharecropping in which the proprietor of the land, who normally
resided in the town, and the head of a farming family made an arrangement to plant a fund (the smallholding) as well as to divide the harvest
roughly in half. Although the creation of smallholdings was also com-
patible with the system of rent—which provided a fixed currency for the
cultivator—and, on the other hand, there were forms of sharecropping
which were not connected to mixed cultivation, between the former and
the latter there was an almost unambiguous relation. The reason is clear:
cultivation had to satisfy, above all, the nutritional needs of two family
nuclei, that of the farmer and that of the proprietor, the first a peasant
and the second a citizen. In any case, it ignored the forms of productive
specialization which were later imposed and which are normal today.
Moreover, the smallholding was also only medium-sized, linked to the
working capacity of a group of workers lacking any mechanical equip-
ment, except in the most recent times. Only what was not consumed by
the double family taking reached the market, ending normally in the
local market that one day a week invaded the main square of the town.
The scarcity of food supplies destined to trade was not unrelated to the
small dimensions of the inhabited centres that constitute the bone structure of central-northern Italy, since it was not possible to feed a more
sustained demographic growth with such quantity of stock. In this way,
there existed a structural link, even if not evident, between the size and

the format of the centres, the landscape and the agricultural relations of production. The essence of the piantata consisted precisely in this close interconnection between urban and rural existence, in their reciprocal interdependence. Its disappearance, almost totally achieved today, reminds us, above all, of the interruption of such a link, and of the division of that age-old and distinctive solidarity.

31. *The Limits of Landscape and the Art of the Actor*

It warns us, rather, that this intimate coexistence was transformed, in the second half of the twentieth century, in the collapse of the single system. To be precise, the limits of landscape as a cognitive instrument consist, for Gambi, in the impossibility of realizing any system. His argument lists a sequence of facts that contribute in a non-incidental or secondary way to the establishment of agricultural realities, but whose reduction in terms of landscape (that is, to something external or something that appears blatantly to the physical senses) is impossible. This is because these facts, leaving conspicuous traces in what is seen, cannot be deduced in a precise and significant manner unless they are previously known, for they are the consequence of what is not seen, that is, mental structures or social institutions. For example, the influence of religious life on the design of fields and roads, often oriented according to the requirements of solar cults, conforming to the compass points. It would suffice, in this regard, to look at the case of Roman centuriation, the systematic, minute colonial layout of roads and canals in perpendicular form, often not following topographic data at all, the configuration of reliefs or the course of rivers. Or else, psychological facts, such as the force of tradition or the custom of imitation, which often determine the manner of inhabiting a place or the type of agricultural practices. Or else, again, the existence or not of the free market, the conflict, that is, between medieval distribution, through egalitarian systems, of common lands (as previously mentioned), and the selective capitalist logic hinged on competition and individual initiative. Or, yet again, the difference between juridical customs relating to family property, on the basis of which, we might note, in the Alto-Adige Mountains, goods are transmitted at the death of the head of the family to the eldest son while elsewhere these are divided among all the descendants, with the consequent fragmentation of holdings. The same agricultural-economic orientations find only a vague and superficial confirmation in visual observation. One can only say that, confronted with the economy of direct agricultural consumption (Slicher van Bath 1960), the economy of the market leads to a greater uniformity in the appearance of the fields, because the number of cultivations tends generally to diminish and plots of land are usually greater in size and more merged with one another.

We might think again, in this respect, of the influence of the city on the countryside, and return to the example of the piantata. We might turn our attention, descending along the Adriatic, to the fare of the vineyard tree, the tree married to the vine, as it was described in agronomical texts of the nineteenth century. On the hills of Le Marche, which incline towards the floodplains of the Tronto, the historical boundary between the State of the Church and the Kingdom of Naples, the piantata still marks its stride in systematized vertical ploughing, that is, along the slope and vertically in relation to the contour lines. Across the river, on the first hills of Abruzzo, the sharecropping enterprise already assumes the typical southern designation of 'farmhouse' and, on the field, the vine and the olive work for themselves, with all direct agreement between the shrub and the tree abruptly disappearing. This is a sign of the weaker and later intervention of the urban capitals, and the poor ability for projection of the city, here historically incapable of impressing the same rule outside of itself which governs its internal order: the syntax of its straight roads (Farinelli 1976: 631, table 98).

That which can be represented of the world on the map, that which is topographically relevant, is not, in short, what can explain its functioning. Essentially, Gambi's critique of the concept of landscape can be synthesized in this observation. The saying of a famous theatre actor comes to mind, for whom searching for life meant finding forms but searching for forms meant finding death. The juxtaposition of the art of the actor and the knowledge of the geographer is not random. Fundamentally, the actor and the geographer both have a relationship to tables. For the actor, these are the stage; for the geographer, they are what we call maps but which were called tables until the nineteenth century. The only difference is that the actor treads upon them with his or her feet and thus subjects them; the geographer treats them instead with gloves and thus, if he or she is not careful, falls prey to them.

32. *Town, Landscape, Dollar Standard: The End of Order*

Modernity ends with the final crisis of mixed cultivation and agricultural production for subsistence. That is, faith in the possibility of the reduction of the world to image ends, and the period of the dematerialization of reality begins. In August 1971, the president of the United States suspended the convertibility of the dollar into gold, declaring the end of the system set up by the International Monetary Fund after the Second World War and ushering in the season of flexible exchange rates (Eichengreen 1996: 174–81). A few weeks later, a law was enacted in Italy abolishing sharecropping agreements which, by transforming sharecroppers into agricultural workers, put an end to what Aymard defined as the 'Italian anomaly' (§ 30). These are two aspects of the same process by virtue of which the activity of the world no longer depends, apart from minimally, on what we can see or touch (§ 24): functions whose results were in perfect agreement prior to modernity (Ivins 1985).

For the farmer, the smallholding was the world. This statement is to be understood not only in the sense of the limited and restricted character of the field in question, but also in the sense of a specific modality of relation to things. This was based precisely on the faith that the relation between the bodily senses and the world was exhaustive, able to define and include the world in every aspect related to the reproduction of earthly reality. The same faith was essentially the basis for the idea that the value of the currency of the greatest economic world power still somehow mirrored gold, the precious metal par excellence. At the beginning of the 1930s, the definitive abandonment of the gold standard signalled the impossibility of continuing to define different individual currencies in terms of determinate quantities of pure gold. Something of the relation between the weight of gold and the value of money still survived, however, in the mechanism of the dollar standard, even if in mediated form. This mechanism still assures the translation between what exists concretely (gold) and what exists abstractly (the nominal value of a banknote), explaining the latter using the former. Today it is very difficult to establish what the value of a currency is based on. One theory makes it depend on the amount of information believed to be controlled by entities (governmental or inter-governmental) which proceeds to be released (Goldfinger 1986: 239–402). What is certain is that a standard no longer exists. Even before this, indeed, there no longer existed something that supplies the concrete measure of abstraction (that is, the value of a currency) with its materiality, serving as a tangible and visible foundation. Landscape and dollar standard, the last reflections of the gold standard, underwent the same process of vanishing, losing their relevance and consistency at once, between the 1960s and 70s (§ 90).

Landscape is a totality which can be attributed to the relation between city and country, if understood as group of visible forms. For this reason, 'village' (*paese*) also loses meaning as the very mechanism which operates at the level of the relation between urban and rural environment. Its greatest and most concise description opens the work on Henri Desplanques' *Umbrian Countryside*, the largest and subtlest study of Italian agrarian geography: 'The plains with their cultivation of trees, the hills with their little olives and acropolises, the mountains with their naked fields, meadows and copses'. He ends: 'There is always a town on the horizon of the worker in the fields' (1969: 1). This description, referring to central Italy, is a sweeping description of the Mediterranean landscape in general. It exemplifies the context this landscape simultaneously expressed and was defined from. Wheat, the vine and the olive make up the nutritional trilogy of the Mediterranean. Mixed cultivation transposed this trilogy onto different but systematically ordered levels. Landscape, in turn, as is shown by Desplanques's sketch, reproduces and expands the motif of tri-partition, confirming this order as well as its visible nature. It is just this consistent compliance of all the things of the world with a single visible order—a model which is cultural even before being material—which constituted what was once called a village (*paese*).

33. *The Village Is Not a Globe, the Globe Is Not a Village*

In the last use of the word, *paese* is used to mean village. Its inhabitants, following what has already been said, make up not only an audio-tactile community, as Marshall McLuhan (1962; It. trans.: 68) defined regions with low levels of literacy. It was also an audio-tactile-visual community, depending on sight in the same way as touch and hearing. If this is true, what McLuhan called the 'planetary' or 'global village' (1989; It. trans.: 59–60) does not exist, despite the success of the formula, first of all because the separation that would fuse it together does not exist. The fact remains, however, extremely significant that the expression dates back precisely to the 1960s and presupposes the loss of meaning in what is seen in favour of what is heard.

According to McLuhan, there is a clear opposition between visual space and acoustic space; although they are in fact inseparable—they are complementary but incommensurable. The first is created by a growth in phonetic Greek literacy which transformed the word, and thus the conception of the world, into something visible, linear, segmented, homogenous and static. The second, arising from the debris of literate civilization as a product of the arrival of the electronic means of communication (radio, television) and electronic technology, would reassemble the total sonorous field of simultaneous relations in new form, which had been the realm of pre-literate man. McLuhan heard the electromagnetic 'tribal drums' roll, which at once delimited and unified the basis of our existence (1962; It. trans.: 39, 62, 68, 174, 14). By virtue of the electrical expansion of the senses, acoustic space appeared to him as possessed not of a single centre but of a plurality of ubiquitous centres. Thus, it was apparently chaotic and in continual flux. Reflecting on this last aspect might help us to understand that the expression 'global village' is a bad metaphor but a suggestive and fortunate one.

We might note meanwhile that a village often has a single centre, precisely the opposite of the globe (§ 8). As to the idea of the Earth as a single very large agglomeration of people, it is not new—it dates back, as Plutarch attests, at least to Stoicism. As such, what remains of the metaphor in the end is not its novelty or its precision but, paradoxically, exactly the opposite of what is normally meant: the confirmation of the reduction of the world to space (§§ 0, 3, 4). This serves to translate the globe into spatial terms; it is one of the many (usually unconscious) conceptual versions of Ptolemaic projection. For McLuhan, and still more clearly for his followers, the village in fact signifies the minimization of interpersonal distance and consequently the maximization of communication. It is, therefore, literally and implicitly transformed into space which, as we know, implies the standard. It is towards the standard, then, that the globe is surreptitiously brought back, in spite of its multiple fluctuating centres.

Better: it is precisely the unconsidered recourse to an implicitly spatial model in the metaphor of the village which stops McLuhan truly reckoning with the logic of globality. It is false that everyone communicates with everyone, that is, to quote McLuhan, that we live 'in an auditory world of simultaneous events and over-all awareness' (1962: 28–9). This is corroborated by the figures supplied by the International Telecommunication Union, the agency of the United Nations which is responsible for issues such as the spread of communication technologies. At the outset of our millennium, 70 per cent of Internet users are concentrated in the highest-income states, where only a fifth of the world population live, along with 60 per cent of mobile-phone users and the same percentage of landline appliances. In the lowest-income states, instead, where double this population lives, landlines do not arrive at a tenth of the world total, and use of the Internet and mobile phones is negligible. One might think that a rebalancing is only a question of time. However, it does not seem so. On the contrary, the gap between the states with the highest incomes and those with the lowest is growing exponentially. If we are surprised by this reality, it is only because we are still used to thinking, following the scientific theory of communication (Shannon and Weaver 1949), that communication is a sequential, logical and linear process, like a mathematical function.

34. *We Inhabit the World, Not Language*

Communication is not, in short, even a question of space, which, from the point of view of possibilities for analysis, if not of the state of the world, is worse. We might think of what occurs in any block of flats in any city, of how little one communicates with next-door neighbours, with people who are physically quite close, even sometimes for a very long period. Incidentally, as was noted by William Bunge (1969: 3), it is not a coincidence that the geography of any single skyscraper is much less researched than the geography of the farthest and tiniest village of the Tierra del Fuego or Lapland, often the subject, instead, of detailed and thorough research.

This happens because, in a sense, McLuhan was much more correct than was believed, but, at the moment of his prognosis, the process of the dematerialization of the world was only at the beginning, remaining thus largely unforeseeable not only in its developments but also in its nature. It is true that acoustic space (different from the space of the village) cannot support any hierarchy, unless in an extremely provisory and intuitively ephemeral way. However, it is also true that the informatization of space proceeds according to processes of very selective channelling, on the basis of largely discretionary aggregations, and thus in virtue of mechanisms which are, above all, exclusive and, in any case, individualistic. The expression 'urban landscape', which is often read and heard, is of dubious significance. This does not detract from the fact that the differences between one urban scene and another can be extremely varied, or that between two cityscapes. One of the most recent is made up, in Italian cities too, of the sequences of satellite dishes emerging out of windows and balconies in the suburbs, especially in the areas where immigrants set up home, or on the edge of the city, or in the countryside. At first sight, this concerns the confirmation of what in philosophy is still believed, that we inhabit not the world but language: in the sense that the world, that is, 'the always-changing circle of decisions and works, of actions and responsibilities', of behaviours, depends on the 'more originary' fact of language: the possibility of assigning names to establish the essence of things (Heidegger 1981; It. trans.: 46). In other and more general terms: it depends on the complex of cultural values that find their vehicle and concise form of expression in language. In this way, the satellite dish, which allows the immigrant to receive transmissions in his or her mother tongue, would signal the attachment to the original form of this possibility, which for Heidegger is 'the foundation of history' and the 'supreme event of being human'. Faced with the not-infrequent case of rural habitations connected to the rest of the world only by a thin trail through the fields, although equipped with the flashy tool which connects them to satellites, one might think that McLuhan and Heidegger, each for his own part as well as taken together, had a point.

If we think like this, we forget the most important thing: that before being a resident, the subject who dwells there is a migrant, and his or her being is the result of an event that implies a displacement, the collision between space and different places. We think like this because we still look at and thus think of the world according to the rules of modern perspective (§§ 4, 5, 9), on the basis of the gaze of a subject condemned to immobility, paralysed as though poisoned by the venom of an arrow, to follow the figurative expression of Pavel Florenskij (1967; It. trans.: 83). The looked-at subject, the immigrant, assumes thus the same characteristics as the looking subject. Looking means, in this sense, to extend one's own characteristics towards the other, projecting one's own nature onto the other. In geography, the same thing that Mandelbrot (§ 10) complained of in mathematics has occurred: the model has taken the upper hand over reality. In this case too, the model consists in Euclidean geometry, here called forth no longer because of the continuity of its objects but for the stasis of its elements. The subject is nothing other than the point (Natoli 1996: 305), since it is immobile and it cannot occupy two positions at the same time. Therefore, to maintain that we inhabit a language and not the world is another way of saying, in the end, that we inhabit not the world but a table—or a map.

35. *The Lesson of Genetics*

To have an idea today of the first movements of our species (§ 52), we might turn to genetic geography, based on the frequencies of deoxyribonucleic acid (DNA) that specify the constitution of the protein molecules of human bodies. On the basis of what is known, these migrations seem to begin just before 100,000 years ago, with the appearance in eastern and southern Africa of the *Homo sapiens sapiens*, that is, the human being who can be defined as modern from the anatomical point of view (§ 53). The most interesting results of research on human genes are, insofar as they interest us, at least two. The first concerns the regular decrease of the genetic resemblance of individuals with the growth of geographic distance. The second consists in the discovery of a high level of correlation, on the global scale, between genetic evolution and the linguistic evolution of populations, and this notwithstanding the fact that languages evolve much faster than genes: owing to progressive differentiation, two idioms can become incomprehensible to one another in less than a thousand years. Evidently, like tracts of DNA, language is transmitted hereditarily from parents to children (so-called vertical transmission), and not only in traditional societies (Cavalli-Sforza et al. 1994; It. trans.: 192, 230–1, 193).

The continent which is distinguished by the greatest amount of genetic data is Europe, which is, however, also the most difficult to analyse in this respect, owing to the complications of its history. Guido Barbujani and Robert R. Sokal (1990) have identified within it 33 barriers, understood as boundaries between different areas, each characterized by a relatively low level of internal variation. According to their study, the whole continent would be subdivided into less than around 40 regions if roughly identified by genetic and linguistic boundaries at the same time (the coincidence of these occurs in 31 cases out of 33). It is said that this coincidence can be explained, in general, by the influence of language on genes rather than the contrary: it would be linguistic barriers which reinforce genetic isolation, and not vice versa (Cavalli-Sforza et al. 1994; It. trans.: 191). This would seem, at first sight, to really confirm, on the scientific plane, the precedence of language over the world. But things are precisely the opposite. It is true that linguistic barriers can be opposed to mixing between different peoples. However, the same two authors are constrained to recognize the essentially physical nature of these barriers, re-prioritizing in this way the Earth and its features over language: 18 linguistic-genetic barriers coincide with seas, or tight straits or fathoms of sea, and 4 with mountain ranges. To confirm the antecedence of the world (§ 0), properly speaking, over language (and over genes), we might look at the remaining 9 cases, in which the boundary does not coincide with any visible physical element but is the product of complex processes within the political and historical-social order. Three of the boundaries in question separate the Lapps from the non-Lapps, distinguishing thus northern Finland from Sweden, central Finland and the Kola peninsula. In addition to a significant original genetic diversity, great cultural and economic differences are involved here, causes and at once effects of some millennia of separated history: all the impalpable elements which hindered, but did not stop, mixed marriages between Lapps and neighbouring peoples. In the remaining cases, religious and political factors are reflected even more immediately, in their historical course, on the geography of genes, elevating frontiers which are invisible to the naked eye which may or may not coincide with those that can be seen. The first kind includes genetic barriers between Holland and Germany, Austria and Hungary, and northern and southern Yugoslavia. The second includes those between northern and southern Germany where the same language is spoken.

At present, a clear distinction cannot be made between factors which contribute to genetic isolation and to the confluence of genetic and linguistic evolution. This could seem a setback, but, if turned back onto genetics, it is instead progress: the most critical and reflective scientists today exclude univocal explanations, refusing to recognize the simple and sole cause of all human behaviour in the precise chemical composition of DNA.

36. *The Illusion of the Genome*

Many others think very differently. It seems that every cell of our body contains within its nucleus two copies of DNA. One of these comes from the father, the other from the mother, and their union derives from the original union of the sperm with the egg. DNA is a very long molecule, differentiated in turn within itself in segments characterized by separate functions which are termed genes. The set of all the genes of an individual assumes the collective name of genome. For many molecular biologists, our body and our mind are the exclusive product of our genome. The infinite variety of members of a species would be nothing other than the result of the different combinations between the various possible genes. As Erwin Schrödinger wrote—among the first—genes are at once 'legal code and executive power', that is, 'the project of the architect as well as of able builders' (1944; It. trans.: 45). We must understand all of this literally. Already at the beginning of the 1990s, DNA was considered a carrier of information which was read by the cells in the course of the reproductive process, in the same way as a series of instructions. The gene was, in short, likened to a living map endowed with the capacity to construct itself as a body. In 1957, Francis Crick stated, with the sequence hypothesis, the rule of this transformation: the specificity of a piece of DNA is expressed solely by the sequence of its bases, and this sequence is a code of the sequence of amino-acids of a specified protein. The task of discovering the code of passage of one sequence to another is entrusted to biochemistry (Keller 2000; It. trans.: 43). The hope was to recover from the map of these sequences (the units of instruction, the messages) useful information for the explanation of all processes of bio-logical variation, starting from illnesses which are reduced to deviations from the normal pattern of the frequencies themselves.

It was an illusion that the recent developments of genetics decisively disowned. The gene whose mutation causes cystic fibrosis was located, isolated and sequenced, and the protein which is codified by it was individuated. However, this resembles many other proteins which make up the structure of the cell, so it becomes difficult to proceed. Con-versely, the gene whose mutation causes Huntingdon's disease has until now eluded any exact localization, and it has not been possible to indicate any defect of a biochemical or metabolic nature as agent of this terrible illness that devastates the central nervous system. Even more problematic is what happens with DNA messages: the same sequence of signals can codify several different instructions, in the sense that all its messages have a function but the function does not coincide the same meaning (Lewontin 1992: 35). For almost half a century, we blindly believed that de-codifying the message in the sequence of nucleotides of DNA would have revealed the secret of life, or would have allowed us to understand the programme which makes an organism what it is.

Already 10 years ago, biologists explained that 'a sequence of three thousand bases can be put onto a CD, and it was therefore possible to slip it out of your pocket and say: "Voila: a human being: this is what I am"' (Gilbert 1992: 94). This illusion rested on the certainty that the structure, material composition and function of a gene coincided in one single physical reality. It was discovered instead that its function is not at all mapped onto the structure, and it is absolutely not possible to assign it a determinate localization within the chromosome. From being a relatively stable and immutable container of linear data, DNA has become something which varies, moving and changing position continually. There are a considerable number of geneticists who have begun today to think that, unlike the chromosome, the gene is not a material object but a simple concept, a by-now-surpassed model. And rather than a map, the sequencing of the human genome is now compared to a yet-undeciphered set of hieroglyphs (Keller 2000; It. trans.: 8–9, 25, 55–6, 80). What was said by Ritter (§ 8) comes back to mind, just as what was said before him by Alexander Pope: searching for life in something which has been torn to pieces, which has been priorly dissected, is a labour in vain because it means losing life in the same moment one believes to find it.

37. *The Violence of Mapping*

The illusion of the genome depends on the violence of mapping. Not for nothing was Anaximander, the first who attempted to draw the boundaries of the *ecumene*—the first geographic image (§ 13)—accused by his contemporaries of impiety, of hubris, of having passed the limit permitted to mortals. He is taught that this is because he allowed himself to represent the Earth and the sea from above, which is only permitted to the gods but not to men. This is sustained but it is not at all true, or, at least, it is not only like this. The true reason for the accusation is another: with his drawing, Anaximander had paralysed and thus killed something (*physis*, nature) that is instead in continual growth and movement. It is the 'genesis of growing things', as Aristotle also notes, a dynamic process and not its inert result. Nature is life, is Dionysus before the Titan's attack, just as the map is his body recomposed by Apollo (§ 2). The CD which the biologist above claimed was equivalent to the human being is thus the ultimate, the most powerful, version of Anaximander's table invented, according to tradition, more than 2,500 years earlier. The principle is exactly the same: making a map of something implies, in Anaximander's language, the prior reduction of a thing to the appearance of the *thing-which-is*, thus its transformation into a being already possessing, by definition, all cartographic attributes, already reduced to a table in advance. The original and silent (because implicit) violence of mapping consists in this very same preventive, and often inadvertent, reduction.

In elementary mathematical terms, mapping is meant precisely as the correspondence between two sets which assigns to each element of the first set a counterpart in the second (Fauconnier 1997: 1). But this definition has the flaw of taking for granted the presuppositions of the correspondence in question, which can be summarized as a 'double faith', the same faith that all maps are based on. That is, that one can assign a precise place to each thing, a definite and specific location, and that every word (every sequence in the case of the genome) has one and one meaning only. We have already seen that in the case of the genome, it is not entirely so. A sequence of code is made up of the initials of the four bases of which each DNA molecule is composed: adenine, thymine, guanine and cytosine. The sequence GTAAGT, for example, is read by the cell in various ways. It is a sort of polyvalent message: it corresponds to the instruction to insert some specific amino-acids into a protein, or signals the end of the message itself. Otherwise, its function can just be periphrastic (that is, auxiliary), or, again, it serves simply to space out one part of the message from another. Neither do we know anything of how the cell decides to read each time. In the language of DNA, as in any complex language, the same terms change meaning with the change of context and change function within the same context (Lewontin 1992: 35).

However, exactly the opposite occurs on a map. We would not know what to do with a map in which names and things were not connected by a bijective relation, in which a name did not refer to one and one thing only and, vice versa, each thing was not referred back to a single name. On a map, all terms are proper names ('Ganges') or, at the most, their specifications ('River Ganges'). Their meaning does not depend absolutely on context, but is given once for all and for everyone, and counts in all situations. 'If a country is triangulated,' wrote Thomas Holdich, one of the officials of the mapping services of the British Empire in the late nineteenth century, 'there is no longer any need for inexact definitions, for confusion about the names of places, and no risk of future controversies.' Conversely, 'every object of some importance receives a designation whose correctness can be demonstrated as easily in an office in London as in a field expedition' (1902: 417). The reason for the accusation of Anaximander can then be thoroughly understood: the map does not only kill the Earth but also mortifies language because it rigidifies not only the object but also the means of referring to it, hence paralysing the subject too.

38. *Squanto*

However, to agree with Wittgenstein, 'What would happen if images (that is, maps) began to oscillate?' (1956; It. trans.: 183). It is exactly this question which has been asked in the past few years in geography as well as in genetics and anthropology: cartographic representations have taken to oscillating, losing, that is, their dictatorial power (as Ritter expressed it) over knowledge and, consequently, the subject, no longer paralysed, has reacquired its freedom of movement, exactly like the gene. Today we are constrained to admit that the gene has no fixity: its existence depends on the dynamic process of the entire organism, and its function within the context of a specific programme of development, capable of changing its own structure in the course of development itself. Likewise, anthropology tends more and more, in our days, to consider the traditional rural village as the waiting room of an airport: as retaining, that is, displacement, the journey, migration not as a simple accident of human existence, as a fortuitous and secondary event, but, on the contrary, as the constitutive practice of the identity of the subject, even of all cultural expression.

An emblem of this turn is the story of Squanto, as told by James Clifford (1997; It. trans.: 29). Fifty miles from the coast, the pilgrim fathers already noted, in November 1620, the smell of the pine trees of the New World. But when they disembarked from the *Mayflower* to found the first permanent English colony in America in Plymouth, Massachusetts, they found on the beach Squanto, a member of the Patuxet tribe who spoke English perfectly because he had just returned from the British Isles. Without the help of Squanto, it would have been much more difficult for the colonists to get through their first, extremely cold American winter. This does not detract from the fact that their first reaction was extreme disconcertment at meeting an indigenous person who was so strangely familiar, and who, precisely because of this familiarity, was paradoxically even more disquieting, since he was so absolutely, unexpectedly different, more than had been imagined: the greatest otherness unexpectedly assumes some features, if not the form, of what we instead believe distinguishes us. The lesson which scholars of cultural processes recover from Squanto's story is condensed in the questioning of strategies that designate non-Western people as 'natives'. This last definition is decisively rejected by anthropology today, which considers the subject no longer as the expression of a delimited field but, rather, of a series of changeable localizations, no longer produced by a static situation but a dynamic process. The subject is an actor, in short, who is no longer internal to a field circumscribed by a frontier but, rather, a contact zone which is quite spread out, made up of relations, interactions and temporary and interconnected behaviours, often based on radically asymmetric, that is, unequal, relations of power, and on fluid and mobile boundaries.

For Arjun Appadurai (1988: 39), natives, the indigenous, have never truly existed, if by native we mean a human being confined in the (and by the) place in which he or she finds him or herself, and not contaminated by material exchanges and exchanges of ideals with the rest of humanity. This conception would be the result of what he calls 'metonymic freezing', in which a part or an aspect of the life of the subject (in this case, the static condition) is exchanged for totality, and ends up marking its conceptualization. In this respect, the same fate befell both the gene and the subject, and at the hands of the same agent: the cartographic image. Appadurai's 'freezing' is the exact equivalent of the process which begins with Anaximander and that, in the last paragraph, was described as the mortification of the object as well as the subject, thus defining the cognitive process. This becomes hegemonic with the success of modern perspective (§§ 4, 34) and only within Humboldtian strategy (§§ 19–22) does the subject reacquire an ephemeral freedom of movement. The traveller stops only to pose himself the problem of representation (§ 25). But if 1848, the year in which the Prussian bourgeoisie came to power, is the year of revolution, then 1849 is the 'year of reaction', Humboldt maintained (Beck 1961: 195–6). It signals, then, the end of the journey and the return to epistemological paralysis.

39. *Knowledge, Recognition, Method: The Geographer's Image*

The subject of the *Erdkunde* was the 'moral man', an individual who sought to observe a rule of conduct that he had given himself, as the expression is correctly translated in a recent version (Ritter 1974: 41). For Ritter, the fundamental rule of scientific work consisted of 'proceeding from observation to observation, and not through opinions and hypotheses to observations' (1852: 26–7). This is exactly how Richard Hartshorne remembers it (1939: 28), in the work which, at the end of the 1930s, defined the form, still today unchanged, of historical conscience in anglophone geography. But Hartshorne omits, in this respect, what is true for Ritter even before this: that 'to be methodical and lead to a natural system, the order of all assembled facts must have a background ideal' (1852: 26–7). That is, it must depend on a hypothesis or on a prior theory (§ 1): only in this way 'can the empirical data be brought into connection, and multiplicity be conducted back towards unity'. Such that, Ritter concludes: 'precisely the most firm conviction that one proceeds without the aid of such a theory in the work of research is in fact the first theory' (ibid.). Or rather: 'the lack of a declared theory does not at all lead more swiftly towards the truth, just as it is extremely far from impartiality' (ibid.). It follows that for Ritter, as for Humboldt, there is never 'simple independent knowledge' (*Kenntniss*) of the Earth's surface, but only 'recognition' (*Er-kenntniss*) of it (Daniel 1862: 18), exactly as, at the dawn of modernity, Christopher Columbus had first demonstrated. With the—single but crucial—difference that Columbus took cartographic representation as his model (§ 7). The followers of the *Erdkunde* were based, instead, on the contrary (§§ 8, 36), on the 'entire content of all valid truths', on the intimate and total vision elaborated by the scientist during the course of his own life 'in Nature and in the world of men', to follow Ritter again. From here, we derive the definition of the *Erdkunde* as 'a discourse too complicated to be able to be encompassed by just any cartography' (Lüdde 1849: *xi*).

However, it is sufficient in this respect to glance over to iconography, to the geographer's image. Already on the eve of the French Revolution, maps were always accompanied by the portrait of the geographer, as a figure immediately recognizable to the viewer. In the early nineteenth century, suddenly, this figure is supplanted by another. The indoor, closed study is replaced by the outdoors and the map by the pen. There are no longer almost any geographers represented in the act of consulting or drawing cartographic depictions, or simply alongside them. Conversely, almost all are portrayed as they observe nature, pen in hand, or while they write. The open air, the natural scenery, implies precisely a journey, movement, the mobility of the subject: all things that did not exist before, in the office of the court geographer. The pen signifies, instead, discourse, and thus the geographer's imperfection, his or her programmatically provisory and partial character, his or her questionable nature completely opposed to the normative and apodictic character of the cartographic tract, which permits neither replica nor criticism (Farinelli 1989b: 45–6). For Ritter, in fact, the *Erdkunde* was only the 'knowledge of the planet which we have thus far acquired as historically determined individuals' (*Erdwissenschaft*). And this because for Ritter and for Humboldt, theory precedes and governs the entire epistemological path and, like the following path, the procedure which is really only re-cognition, it functions on one condition only: the prior admission of the historically and socially determined, and thus relative, character of one's own nature. Therefore, Ritter defines the Earth as 'the greatest of living individuals' (Daniel 1862: 17, 32), assigning to the planet and to the continents the same form as the subject of geographic knowledge. The representative of 'civil society', still antithetical to the aristocratic-feudal state, is agent of the 'bourgeois public sphere', who, precisely in these years, was conquering, on the juridical plane and as protagonist of the market, its personal emancipation, the right to its subjective singularity, namely, to its individuality (Farinelli 1981a: 53).

40. *The First Death of the Master of Ballantrae*

This subject suffers the same fate as befalls the Master of Ballantrae in Robert Louis Stevenson's eponymous novel. As everyone knows, the Master of Ballantrae fakes his death three times—he escapes the jaws of true death three times by pretending each time to really be dead. In reality, he is only momentarily paralysed. The same counts for the subject of geographic knowledge.

Its apparent first death coincides, between the nineteenth and the twentieth centuries, with the invention on the part of Paul Vidal de la Blache (1922) of 'human geography', an adaptation to the French culture and taste of Friedrich Ratzel's anthropogeography. The original meaning of Vidal's lesson, founder of the anthropogenic kind of geography hegemonic in all of the twentieth century (Pinchemel 1972), was based on 'an urgent invitation to observe the map', as his contemporaries noted, not without surprise (Bourgeois 1920: 19). Surprise depended on the fact that this invitation was understood as the abandonment of every form of 'memory', that is (as Hegel taught), of every idea of the existence of consciousness and language. It must be recalled, by contrast, how it was philosophy, history and language that were instead the armature of the *Erdkunde*. Paradoxically, Vidal's training was historical at the same time as being archaeological. But after the unforeseen defeat at the Battle of Sedan, after the unexpected and humiliating conclusion of the Franco-Prussian War, he was literally constrained by the government of Paris to improvise himself as a geographer and to found a modern national school of geography; the inadequate knowledge of the French territory on the part of the army was identified as one of the causes of the defeat, and it was necessary to rectify this (Berdoulay 1981: 30). Hence, in cartographic representation, Vidal saw 'the instrument of precision, the exact document which straightens false notions' (1904: 120). This may work for making war, but it is a little less useful if one tries to apply it, following Vidal's intention, to scientific discourse. As even today it is taught in universities across France, he 'started always from the real, avoiding all which smacks of theory or a priori construction. He had learnt to adhere to the concrete, that is, to the map' (Claval 1964; It. trans.: 69). This expression, which implies approval without reserve, demonstrates two things. The first is Vidal's long-lasting influence, which endured in Europe (except in Germanic countries) throughout the twentieth century. The second concerns the nature of human geography, which in this same refusal of all prior theory, and in the immediate adhesion to the cartographic dictate—understood as a synonym for reality—is configured as the exact reverse of the preceding German critical geography of the *Erdkunde*.

What follows is the (apparent) death of the subject of geographic knowledge, its paralysis, its immobility: every critical epistemological journey, that is, every procedure that starts from a theory based on a relative conception of the world, arrives at a final scientific description—it is abolished insofar as it is superfluous, in the moment in which there is no longer any distance between the former and the latter. And there is no longer any distance because, from the outset, for Vidal and his followers, there exists a description—the one represented by the cartographic image—which in its real concreteness (or in its concrete reality) allows one to do without any theory because it is already scientific by definition. 'To describe, define and classify, therefore to deduce' (Vidal de la Blache 1913: 298): this was for Vidal the correct 'methodological succession', where for 'to describe' we must understand to make a map or to infer the nature of things from the map, in the naive sense of a restoration of reality lacking any theory (thus a priori untainted). With Vidal, and with positivistic geography, which in the last quarter of the nineteenth century was centred on Germany, the map returns to being the same thing that it was in the time of aristocratic-feudal geography (Farinelli 1992: 107–50, 156–67): a formidable ontological dispositive, a silent instrument for the implicit definition—that is, not subject to reflection—of the nature of the things of the world. Except that against it, neither the ire of contemporaries, as in Anaximander's time (§ 37), nor irony, as in Herodotus' time (*Histories*, IV.36), is thinkable any longer.

41. *What Is a Tree?*

To give an example. For Vidal, the 'instrument of precision' par excellence was the topographic map. In the manuals of cartography, a representation is described as topographic if its scale (§ 3) lies between 1:5,000 and 1:200,000, that is, in which 1 centimetre is equivalent at most to 2 kilometres in reality (Selvini and Guzzetti 1999: 78). It should be reiterated in this regard that the relation expressed by the scale exclusively concerns lengths and does not count for any other dimension. Cartographic representation sacrifices, if necessary, all other features of the world in order to supply just one piece of information with the greatest precision: the linear interval between two points. In any case, from a historical perspective, the topographic map is not just any map: it constitutes the portrait, exclusively entrusted to the military, of modern national territorial centralized states, the image which these have produced of themselves and publicized beginning from 1840 (Stavenhagen 1900: 510–11). Like all maps, the topographic map also functions through an inevitable choice from among the innumerable elements of which reality appears to be composed. Without this choice, every map would be incredibly crowded—one single black mark—and we would not be able to distinguish anything at all: the most faithful representation would correspond to the surface of a board without any graphic expression, completely empty. But this image would be entirely useless. So there arises for each map the problem of selection, and of the process of reduction or modification which concerns the size, shape and number of phenomena represented (Balodis 1988: 71; Neumann 1977; João 1998).

The semiotics of graphics distinguishes in this respect between 'conceptual generalization' and 'structural generalization'. The first corresponds to a 'change of framework' of the phenomenon, and thus a new conceptualization of the latter accompanies it, as when the forest is substituted by a tree. 'Structural generalization', conversely, implies the conservation of the given conceptualization. This is limited to the simplification of its structure (Bertin 1967: 300–01), as when the symbol which stands for the forest passes above, let's say, a number of trees proportional to its size, to a fixed and limited number of trees for every forest. This distinction, however, is limited to illustrating the passage from one scale to another; it says nothing about the original problem of the reduction of reality to cartographic sign. Proof of this is the different treatment reserved for the same single tree on topographic maps at the same scale.

We might take, for example, topographic maps at the largest scale which cover the entire national territory, the 'little tables' at 1:25,000 of the Military Geographical Institute. Like the animals in George Orwell's *Animal Farm*, all trees are equal but some are more equal than others, in the sense that there are two types of trees: those indicated by a generic sign that corresponds to a very small circle (namely, the tree), and those distinguished by a particular grapheme, by a little drawing which reproduces its stylized outline (the oak, the mulberry, the olive tree and so on). Why this disparity of treatment? This does not spare any being in particular but depends on a set of relative criteria. Let's examine the simplest case.

Let's look again at the pattern of the piantata or alberata (§§ 29–30), which is the outcome of a complex historical, social and economic reality, even before that of particular physical and climatic conditions. In the south, it represents an exception. However, only in the Campania plains is it actually indicated with a specific symbol (two poplars linked by a strand of vine) while in all the rest of the peninsula it is represented by the standard alignment of two generic signs that everywhere stand for the vine and the tree. The explanation is simple: in the Aversa countryside, where trees even exceed 20 metres, exuberant natural agents (the climate and the soil) favour their imposing size, and it is this which topographic form is concerned with registering and underlining (§ 63). That is to say, in the case in question, cartographic representation transforms the most artificial and sophisticated rural architecture into a natural product. And precisely in this sense does it act as an authentic ontological dispositive (Farinelli 1976: 631–2, table 99).

42. *The Second Death of the Master of Ballantrae*

In short: everything which is the result of historical and social processes is transformed by topographic representation into a natural formation, into a simple, material aggregate. One might say that it changes nature, if it did not seem mere wordplay. This occurs because, evidently, the cartographic system is an authentic logical system, as Ritter and Humboldt knew very well, and as geographers began again to admit only after the Second World War. Representation on the map is a theory that (more or less consciously) geographers accepted, as E. L. Ullmann (1953: 57) observes. But instead of pushing the analysis of this theory to a criticism of it, the admission of the theoretical character of the cartographic image, unthinkable for the preceding geography, becomes the recognition of a sort of fait accompli, in which nothing remains but to draw out the final consequences. William Bunge will later do this: 'Maps have represented the logical structure on which geographers have constructed geographic theory' (1962: 33). But these, he adds, are not anything other that a subset of mathematics, thus a type of uncompleted restoration of the world in mathematical terms, whose only fault is precisely in its imperfect—because incomplete—character. If geographers, continues Bunge, have up until now trusted maps, they will have to trust mathematics even more, because maps themselves are based on its principles. Why, in short, be satisfied with a partial translation such as cartographic translation, and not proceed to a version of geography that is thoroughly formalized, that is, mathematical?

This was exactly the programme of 'quantitative geography', which dominated the anglophone scene, and was reflected in the European scene, in the 1960s and 70s. Within this, single concrete phenomena are replaced by abstract properties, that is, geometrical properties, of the pattern of their distribution in space, so that a river basin is transformed into a graph of the water supply; the cities of a region assume the shape of a more or less symmetrical polygonal mesh; the apparently chaotic use of soil is put in order, for example, by the design of ideal concentric rings. Simultaneously, the appeal to mathematical language, much more rigorous and abstract than natural language, allows the immediate quantification of empirical phenomena (Burton 1963; Dematteis 1970). According to Brian Berry, the most famous exponent of this trend, it would finally be possible in this way to proceed towards a fundamental distinction: that between 'the *facts* that constitute the object of geography, the *theories* that insert facts within the *models* based on the perception of spatial order, and the *methods* used to link facts to theory, so formulating the latter in the most precise and concise manner possible' (1960: 282). But nothing is said about what a 'fact' or 'theory' is. The nature of the object under investigation is from the outset removed from reflection and determined instead in a mechanical, and thus inadvertent way, by the conditions for the practice of the applied method themselves. For quantitative geography, the fact is, in reality, only what can be reduced to quantity—thus measured—and all which cannot be measured is then excluded from analysis (§ 90).

In this respect, quantitative geography is truly the unconscious continuation of nineteenth-century positivist geography. It eliminates definitively the subject of geographic knowledge since it entirely suppresses the question of meaning, which is to say, the intention assigned to geographic knowledge within the context of the social totality, within the world. For Gregory Bateson, a fact was the result of a description, and a theory was the description of a description. Vidal started again from description, although definition of a cartographic nature and, by definition, unquestionable. Quantitative geography starts instead from the last of the terms of Vidal's methodological process—from deduction. He does not call it by its name but as 'fact', and since this assumes geometrical-mathematical robes, it is claimed that its rigour extends to the entire impersonal mechanism that begins from this secretive deduction. This confuses, then, the precision of form with the correctness of procedure.

43. *What Is Zero?*

Bunge's error lay precisely in the original assumption according to which cartographic representation was a crude derivative of mathematics, an as-yet-insufficiently-refined version of it. The opposite is in fact true: it is not the table (that is, the map) which derives from numbers but it is number which derives from the table. We might take the case of the most elusive, mysterious and dangerous number: zero.

In the accounting system of late Babylonian civilization, the zero— two small wedges, slanting and parallel—were simply a sort of signpost, a symbol for an empty space on the abacus or calculating table. This corresponds to a column in which there was nothing, because all the pebbles (*calculi*) were already piled below. It had neither substance nor intrinsic value; it was a figure, not a number. In the West, the use of the abacus is documented from the seventh century BCE, and is contemporary with the creation of the first geographic representation of the world, Anaximander's table (§ 13). In both cases, we are talking of tables, and the practice of those who calculated was the same as that of the cartographer, consisting in filling the gap presented that was devoid of symbols or signs, a blank space. Exactly like the blank on a map, the zero on a calculating table was not a thing, a number, but a condition, normally transitory and momentary, a part of the table itself. It was Leonardo Fibonacci who would, with his *Liber Abaci* written in 1202, introduce arabic numerals to Europe, including zero. But how had this number become, in the meantime, the most important number in the East?

According to Robert Kaplan (1999; It. trans.: 76), the empty circle which today stands for zero comes from the mark left by the circular pebbles on the surface of a calculating table covered in sand, so as to leave a trace, that is, a memory, of the calculation itself. This passage is decisive for at least two reasons. First, this is exactly the process of projection (§ 4): in both cases, we are in the presence of the transformation of a globe into a two-dimensional sign, subtracting one dimension from the sphere. For Wittgenstein, projection is a way of changing the meaning of something through a change in the technique of its representation (Diamond 1976: 43, 49). And this is also what occurs in the same way in relation to the zero. The second reason we might pose is in fact the following: if the round pebble leaves a mark, this means that after a series of calculations, all the columns in which the table is subdivided bear the sign of zero, or (and this is even more important) they can bear it. This is equivalent to conceding something which should already have become clear but which only the existence of the mark renders obvious: the zero stands for the part of the table that is not occupied by pebbles—thus, very simply, the zero stands for the table itself; it is the table itself.

Precisely for this motive, one might venture that on the Pythagorean chart, zero does not exist. Because Pythagoras and his followers already knew that the table itself was the great nothing from which all notation derives, the first great zero that has no value in itself but that through its existence gives value to every other thing. For Pythagoras, every thing was a number, something, that is, which originated from the table, the formidable instrument that no philosopher, excepting Pythagoras, has ever stooped to reflect on seriously (Kaplan 1999; It. trans.: 40). Between *mensa* and *mens*, between the Latin terms to indicate tables and the mind respectively, there is, after all, an evident affinity, almost a coincidence. This does not mean only that we can compare our mind to a table but also the contrary: that we must consider the table (the zero) as a mind, capable of producing ideas. If this all seems excessive, one can appeal to the words of Mahavira, Indian mathematician of the ninth century CE: 'Zero becomes what is added to it' (ibid.; It. trans.: 104). Words which apply equally well for the table that, joined with individual signs, we call a map: Why should we otherwise believe that a city portrayed on a map is actually that city, if not because the table itself on some level becomes it?

44. *The Third Death of the Master of Ballantrae*

The false deaths of the Master of Ballantrae happen by freezing, and those of the subject of geographic knowledge (also abducted, like the first, by American pirates) by 'metonymic freezing' (§ 38). Based on the statistical treatment of figures and on the probabilistic approach, quantitative geography substitutes the search for the causal link (for the why) of phenomena by the search for their probable line of tendency (for the how). Simultaneously, it tries to infer hypotheses relating to reciprocal connections, or to the interdependence of phenomena themselves, from the greater or lesser level of correlation of numerical figures which result from measurement. But a numerical figure, which represents the starting point, is only a final valuation equipped with the property of no longer containing any trace of the choices and considerations which led to it. That is to say, within quantitative geography, there is no space for any type of subject, in any moment of the analysis. Its function consists, fundamentally, in the application of the order which already no longer functioned in the field of the visible (§ 24), to the invisible structure of processes. This is the order of Euclidean geometry, indistinctly applied to physical phenomena and to historical-social events.

The subject returns to make problems for geographic knowledge at the beginning of the 1970s, with the birth of so-called behavioural geography (Gold 1980). In reality, it is a further form of demise because, more than a true subject, it technically concerns a type of zombie, a sort of living dead. In Voodoo, the national cult of Haiti, a zombie is the soul of the departed, separated from the body and used for magical ends, or otherwise the body without a soul, reduced thus to an automaton without any will, which moves only to obey orders. This is precisely what occurs within behavioural geography. Different from what occurs in quantitative geography, the subject in behavioural geography is represented as a variable interposed between the environment and spatial behaviour. But this latter appears as the result of a simple link between environmental stimuli and reactions, as if social behaviour was not determined by aims and values on whose basis the acting subject orients itself, and which need to be included on pain of a lack of comprehension of behaviour itself. In fact, social processes are not composed of mechanical behaviour or behaviour possessing an objective meaning, but are based on a sense, and thus a motivation, which has a subjective value, built on a subjective interpretation of the facts. Consequently, the same behaviour must be understood by passing through the interpretation of the interpretation of the world on the part of the acting subject, even if this subject escapes immediate observation. It is the principle of 'comprehensive sociology', which results from Weber's great lesson on the theory of social action (1951; It. trans.: 239–307). A lesson which has remained foreign, instead, to the analyses of behavioural geography.

In relation to quantitative geography, this reintroduces a problem which, at least from the end of the nineteenth century, was forgotten in geography: the question of the preliminary hypothesis about the nature of the object under investigation (§ 61)—in this case, man. However, the form of this reintroduction immediately reflects the insufficiency of the operation: man is assumed as simple elaborator of information, and nature as simply a set of data to elaborate. Reality is, in this way, reduced to a system which functions on the basis of the *input–output* model, where the environment is only the set of inputs to the system. To study something means, as such, to examine its output (the change produced on the environment) and its relations with the input (the external events or factors that change the thing itself). Those who accept this way of seeing man and nature are defined by J. David Bolter (1984) as 'Turing men', from the name of the mathematician who in 1936 first supplied the symbolic description of the logical structure of what, within a dozen years, would become the first computer—and which still governs the laptop on which I am writing.

45. *Who Moves and Who Stays Still*

In a text of ancient Chinese wisdom commonly described as the 'Bible' of Taoism, the *Tao Te Ching* (Duyvendak 1953; It. trans.: 170), the happiness of an ideal country is described at the end. There, among other things, work tools are not used, nor are means of transport, men do not emigrate and, even if villages exist which are so close to one another that one can hear the cocks crowing and the dogs barking, their inhabitants never visit other villages. We are talking of a land which has probably never existed, at least in historical times and at least in the West. If Ulysses had not heard the voices of the Cyclops and the bleating of the goats coming from the land in front of the beach where he had landed with his comrades, space would never have been invented, as we will see. It is precisely the land of the Cyclops—neither ploughed nor sown, and inhabited by giants who live by the sea but do not know of boats—which resembles the country described by the Taoist master. The confrontation between Ulysses and Polyphemus is the confrontation between those who know law and assemblies, and who thus act in political terms, presupposing the existence of the city, and those who know nothing of all this. But even before this, there is the confrontation between those who move and those who stay still—the original opposition, whose outcome, favouring mobility, made mobility the fundamental condition for all that we call culture.

If we were to inhabit not the world but only language, then it would be possible to live by staying still. But if we inhabit the world it is much more difficult. The same concept of the *ecumene* (§ 1) presupposes movement in some sense, that is, the extension of knowledge and consequently of the inhabited portion of the world, even if the Ancients had a quite rigid and limited conception, seen with today's eyes, of the inhabitable. According to Erastothenes, the *ecumene* stretched to around 9,000 kilometres in length, from the latitude of the British Isles to Taprobane, the island of Sri Lanka. It was believed that outside this 'temperate' zone, the Earth did not offer the possibility of life to man, because of either excessive heat or excessive cold. Geographers of the last century meant by *ecumene* the group of lands in which man lives permanently and reproduces. Following this definition, in our times, there would remain around a sixth of the globe's surface excluded from the *ecumene*, all the area which is located south of the imaginary line of connection between the lower points of the continents in the southern hemisphere: essentially, the Antarctic continent, the coldest of all, as well as extending below the line which connects the Tierra del Fuego with southern Georgia, the Cape of Good Hope, the Mascarenes, Stewart Island and Easter Island (Ortolani 1992: 77). It must be added that in the past few years, the Antarctic has seen quite intense activity in terms of temporary residence for scientific purposes and, more recently, tourist purposes, so as to suggest a further increase in the near future.

This is the last episode in what was celebrated as the conquest of the Earth between the nineteenth and twentieth centuries, culminating, on the eve of the First World War, with the reaching of the Poles: the American Robert Peary was the first, in 1909, to claim to have reached the North Pole; the Norwegian Roald Amundsen, in 1911, and the British Robert Scott, in 1912, were the first to reach the South Pole. This is certainly not concerned with, in the case of exploration, immediate reactions to environmental stimuli such as behavioural geography would claim, but connected to a complex plot of multiple and nuanced forms of economic, political, social and cultural mediation. And the same counts, although in different measure, for all the other forms of geographic mobility. At the same time, every form of migration is a transfer which moves from one place to another on the surface of the globe according to a certain order. In the middle of the twentieth century, it was Carl Schmitt (1974; It. trans.: 59), reader of the *Erdkunder*, to remind us that the term 'nomad', which indicates those who customarily inhabit the world by moving, derives directly from the Greek *nomos*, that is law: an order inscribed in some way on the Earth itself. Exactly the thing that was dear to Carl Ritter: 'the terrestrial order of our planet'.

46. 'The Terrestrial Order of Our Planet'

With this expression, Ritter referred to the asymmetric relation between 'fluid forms' and 'rigid forms' that govern the surface of the Earth, and to the unequal distribution of sea and land. The asymmetry finds expression foremost in the different extent of their respective areas (361 million square kilometres for the first, 149 for the second), but in fact concerns their inconsistent distribution. Ritter distinguished a hemisphere to the north-east, centred on Europe, and a relatively vaster water hemisphere to the south-east, linked by a chain of islands interrupted only at the stretch between Cape Horn and the Cape of Good Hope: from the southern tip of America to that of southern Africa; from here to the Chinese-Malaysian Mediterranean (Indonesia, Malaysia and the Philippines); and later to the volcanic archipelagos that line the eastern shore of Asia and protrude, beyond Japan, towards the Kamchatka Peninsula, Alaska, the American North-West until reaching, across the Californian Peninsula, the southern point of the American continent again. The opposition between these two 'universes', the land (that is, continental) and the water (that is, Pelagian), constituted for Ritter the fundamental and original contrast of our globe, the one upon which all others depended (Ritter 1852: 104, 206–46). In 1904, Halford Mackinder explained Ritter's vision (never elsewhere referenced) as what can reasonably be called the first geopolitical doctrine and which still enjoys some favour in the major states (Mackinder 1904).

Mackinder distinguished 'three natural seats of world power': an entirely continental area, the 'greatest land mass', the 'heart of the Earth', the interior of Eurasia itself, surrounded by ice to the north and, in any case, inaccessible to ships, which therefore remained outside Europe that was bathed in the Atlantic rains; a internal crescent, a peripheral belt accessible to men of the sea and stretching from western Europe to China and Kamchatka; and an external crescent coinciding with Ritter's water hemisphere. Mackinder's problem, like Ritter's, consisted in considering 'human history as part of the life of the world-organism'. In

Mackinder's vision, the first area, extending from the Arctic coasts to the deserts of Central Asia, and to the west to the wide threshold separating the Baltic Sea from the Black Sea, was precisely 'the geographical pivot of history' and, in strategic terms, the greatest natural fortress in the world. As Mackinder (1943) will warn in the course of the Second World War: if those who occupy the pivotal region (at that time the Soviet Union) can expand into the lands of the internal crescent (at that time Germany) it will become the strongest power in the world, because it will reach access to the se and alter every balance inscribed into the order of the world by nature.

But before Mackinder's launching of geopolitics, Ritter's work establishes, with all due acknowledgement, Hegel's philosophy of history. The latter indicates his debt in relation to the first volume of the *Erdkunde*, dedicated to Africa, where it is claimed that human history follows the course of the sun, moving from east to west (Ritter 1852: 10–15). From Ritter, Hegel takes the idea that the foundation of the historical process is by nature geographic, that is, universal history obeys a development that depends on the different physiognomy of continents, because their configuration conditions the destiny of peoples. While for Ritter the realization of history involves the entire globe, Hegel limits the historical theatre to the temperate zone of the Old World. For Ritter, east and west were relative and mobile terms, which the discovery of the New World had suddenly forced to redefine. For Hegel (1996[1837]), instead, these remain absolute terms: the first coincides with Asia, 'the continent of origins'; the second with Europe, the final half of the historical process, of the definitive realization of the 'world spirit'. In this way, the Euro-African Mediterranean becomes for Hegel the 'axis of universal history', the only geographical centre of the world (Rossi 1975: 41)—something inconceivable for Ritter, for whom the world remained a sphere, thus possessing an infinity of centres.

47. *Mediterranean and mediterraneans*

Until now we have simply said 'Mediterranean', but it has now become necessary to specify which mediterranean we are speaking of. The name Mediterranean, as Pierre Deffontaines wrote, should always use the plural, because it is composed of a group of separate seas, confined basins, which follow from one another from east to west 'like rosary beads' (1972: 13). Deffontaines was both mistaken and correct, for a reason soon to be revealed. Ferdinand Braudel (1949), though, was merely wrong when he underlined, as the motto for his life's work, that in the American continent there exists nothing which can be compared to the Mediterranean. He was mistaken, because a mediterranean is nothing other than a large oceanic gulf identified by a profile within which the mainland prevails decisively over the water element, and there are several of such gulfs—which is why it is truly never necessary, in this respect, to use the singular. There exist many potential mediterraneans, several seas almost completely surrounded by reliefs, more than a number of 'liquid plains which communicate through quite large ports', according to the definition which Braudel (ibid.; It. trans.: 102) reserves to the Mediterranean of the Old World. There also exists at least one of these in the New World, an American mediterranean made up of the Gulf of Mexico and the Caribbean Sea, just as there exists a Chinese-Malaysian mediterranean composed of the southern and eastern Chinese Sea, the Yellow Sea and the Indonesian and Philippine Seas. The other two mediterraneans were until now (that is, historically) singled out as functional units, both as first turn of the Tropic of Cancer and as possessing roughly the same extent: a corridor of approximately 4,000 kilometres in length and about 1,200 kilometres at maximum width.

But pay attention to the compact group formed by the Hudson Strait, from the Hudson Bay and the Foxe Basin in the northern part of the American continent. Or consider, north of the Korean straits, the two large basins, aligned vertically and interrelated, the Sea of Japan and the Sea of Okhotsk. From the physiographic perspective, that is, at first sight, the examples just mentioned are fully included in the category of mediterranean. But why does the image of the first consist in the simple succession of a series of inlets, and the second purely in the series of different seas? In other terms: What transforms a simple internal sea into a true and proper mediterranean? The answer is simple: before all else, the indispensable function of aperture for inter-continental or inter-oceanic communication, as precisely (still?) is not the case for the Hudson Bay or for the seas that separated Japan and the Kamchatka from the Euro-Asiatic continent. And as, instead, occurs without exceptions for the marine stretch between Europe and Africa, that between northern and southern America and that between Asia and the Australian continent. The sole difference is that the lands that make up the *Mare Nostrum* of the Romans are much closer and more attached than one finds elsewhere: the Strait of Gibraltar, which separates Spain from Morocco, is 13 kilometres wide, and the Bosphorus canal, which divides Europe from Asia, is only a kilometre and a half. Conversely, the beaches of Florida and Cuba are separated by almost 200 kilometres, and the distance is almost double that for Formosa and the Philippines, from one point to another in the Luzon Strait, the door between the Pacific and the the South China Sea.

Yves Lacoste is of a different opinion (1982: 5): for him there exist only two mediterraneans, that of the Asiatic south-east not entering the category because, unlike the others, it is not interposed between two great continental blocks. But the interpretation appears rather restrictive and reflects the Western model too greatly. Mediterranean in fact means medium, that is, a means of communication, between lands. We are speaking, thus, of a proper name that directly designates a role: that of an immense 'space-movement', a single system of circulation in which land and sea routes have reached the point of becoming indistinguishable. Precisely by virtue of this function, and complimenting this general territorial equivalence, the Mediterranean today no longer appears singular, and in its own interior less and less diverse (Farinelli 1995).

48. *A Small Quarter-Turn*

For Ptolemy (*Geography*, I.1), geography was the description of a head, and chorography was of one of its parts, for example, the ear. Projection was the system which ensured the equivalence of the collection of parts (of maps) with the whole. Every map and every part has its own centre. Thus the Earth is absolutely a labyrinth for Ptolemy, as it is for Ritter. The difference between the two, that is, between geography and the *Erdkunde*, lies in the fact that while in the first case the group of maps is sufficient to restore the globe, in the second, the opposite occurs: the sum of the parts does not correspond to the whole. This latter remains something irreducible to the sum of its individual components. The geographic vision is additive; that of the *Erdkunde* is, instead, holistic, based on the awareness that even in putting together all the possible cartographic representations of the Earth, there remains something left over: the labyrinthine nature of the world, which in the cartographic version is, instead, negated. This is another consequence of the irreducibility of the sphere to a plane (§ 4). But understanding this entails understanding the behaviour of the subject of knowledge.

In this regard, Ptolemy (*Geography*, I.20) is very clear, even if rather reticent. He recommends making maps and not resorting to the spherical model for practical reasons: the globe is inconvenient to use, he writes, because we must continuously circle around it, or we must continuously slide it around with our hand. In both cases, it will be noted, the subject is condemned to move exactly like those that, without any map, wander inside a labyrinth. Conversely, if the image is a map, the subject has no need to move because it has no need to look for the centre and sees everything immediately (§ 8). We delude ourselves that today we are no longer Ptolemaic, only because we no longer believe that the Earth is actually the centre of the universe, as Ptolemy taught in astronomy. In reality, we are still truly Ptolemaic and we profess our unconscious faith every time we open an atlas, since it was Ptolemy himself who first reduced the world to a series of defined points from a pair of mathematical coordinates calculated through astronomical measurements. Even before this, it was Ptolemy himself, prohibiting the globe, who established definitively that the subject was to remain immobile, that knowledge was to be a function of a double, connected stability: that of the subject and that of the object. It is true that the Western subject who today watches the immigrant projects, insofar as it is a modern subject, its own static nature (§ 34). But, in turn, this nature is derived by the modern subject from the fixed and self-centred nature of its own image of the world, the cartographic one. Even before the advent of modern Florentine perspective, the effect of projection was, already with Ptolemy, that of paralysing the gaze, transforming the human being into a statue, exactly like the Medusa, the monster who retaliated against men, doing to men what men did to other parts of nature, transforming them into things, and thus changing the subjects its eyes met with into inert stone objects. At the beginning of the nineteenth century, even Hegel's philosophy of history falls prey to its deadly charm: the course of the sun stops, the West and the East are blocked (§ 46). Only the *Erdkunde* still resists by considering the Earth a globe. For this reason, Ritter, looking at it sideways and from below, 'from behind' (§ 20) and thus walking around it, can distinguish a water hemisphere from a continental one. Hegel instead considers the globe as though it were a map, centred once and for all on the Mediterranean as in Ptolemy's time. Therefore, anthropologists today enjoy (Clifford 1997; It. trans.: 20–1) asking themselves: For Hegel, the owl of Minerva, that is philosophy, spreads its wings at dusk. But where is the dusk if the Earth revolves? For whom and how many people is it dusk?

Michel Serres notes, astonished, that it takes a quarter-turn of the point of view, from below upwards, for the archaic model of the world, in which the Earth rested on a vase, to coincide with Anaximander's geometric model, product of the projection of the vertical dimension onto a plane (1993; It. trans.: 99–105). What Serres, however, does not note is that this turn implies the mobility of the subject, which, with the invention of the geographic table (§§ 13, 37), was first immobilized by Anaximander himself.

49. *Frogs, Pond, Earth*

For Hegel, as for Plato, mankind remained a group of frogs crouched around the Mediterranean Euro-African pond. Humanity still predominantly inhabits coastal regions around the world pond, following the effect of multiple asymmetries. Some are physical in nature, others result from relationships between the Earth and humanity as organized collectively. The main one, which concerns the unequal subdivision of the Earth's surface between water and land (§ 46), is further accentuated by their different distributions and different forms, depending whether we speak of the Northern or the Southern hemisphere, into which we often divide our planet. In the first, in the Northern hemisphere where the continents tend to spread out towards the Arctic, two-thirds of the landmass is concentrated along with nine-tenths of the approximately 6 billion people who make up today's world population. Heading south to the Southern hemisphere, the landmass shrinks and the human population decreases sharply, with some few dense areas as exceptions: California, the Brazilian coast, the Rio de la Plata estuary, the island of Java, south-eastern Australia. These are incomparably smaller demographic foci, in terms of consistency and extent, in comparison with the Northern ones. With almost 800 inhabitants per square kilometre, Java is certainly the island, among the large islands, which is the most densely populated on the globe, even more than Japanese islands (we might note, to give an idea of the scale, that the average density on the Italian peninsula is 191 inhabitants per square kilometre). But the total number of Javanese people, just over 100 million, amounts only to a seventh of the European population, to a ninth of the Indian population and to a twelfth of the Chinese population, which together constitute half of humanity. In this way, half of the present inhabitants of the Earth are crowded onto a sixth of the land, by a rough calculation, based as it is on a wide selection taken from different political-state bodies. In reality, density varies above all in China and Europe, and varies greatly as we move inland from the coast. If by China we mean, in fact, the historical China of the 18 eastern provinces, where more than a billion inhabitants live on 4 million square kilometres, and if by Europe we mean Mainland Europe, west of the area between the Black Sea and the Baltic Sea, then the Eurasian concentration becomes almost double: more than 2.5 billion people, almost half of the total human population, on just over a twelfth of the land (not even 2.5 per cent of the total surface of the Earth).

A very recent study (Mittermeier et al. 2003) allows one to appreciate the reverse, as it were, of the concentration of demographic pressure. Sixty-eight million square kilometres, thus little less than half of the surface of all the continents and islands, are still in a wild state. This attribute describes all land possessing the following characteristics: more than 70 per cent of its original vegetation is intact; its extent is not less than 10,000 square kilometres; population density is not above five inhabitants per square kilometre. Almost a third of such areas are made up of glaciers, mostly in the Arctic. The Arctic tundra, the cold desert without trees immediately below the North Pole, counts for a quarter. The northern forest belt, which develops in turn south of the tundra, contributes just under a fifth. Following are the hot deserts, with just over a sixth, and tropical rainforests with little more than a twelfth. Five regions are wilder than the others: the Amazon, which possesses more than half the forests of the planet and the highest concentration of living species; the Congo, also rich in forests; New Guinea, which is the richest tropical island in terms of animal and plant life; the lower basin of the Zambezi river in Mozambique, where more than two-thirds of living elephants are found; the hot deserts of South America. These areas, which count for just over 7 per cent of the land, contain more than 60 per cent of the plants and more than 40 per cent of all the vertebrates of the planet. That is to say, the distribution of plant and animal life, as much as that of humans, is concentrated and asymmetrical.

Why should it be, on the contrary, symmetrical, that is, uniform and immune to possible change (Rosen 1995: 1–4)?

50. 'The Crooked Timber of Humanity'

By this Kant meant the innately imperfect nature of humanity, its weakness and infirmity of constitution. How and when, however, was the idea of right born in Western culture, that is, the idea that what is straight is also good, better than what is not straight?

To respond, we must return to Polyphemus' cave, where Ulysses and his comrades are trapped. The problem is getting out, and this requires much more than a trick of names (§ 17). The *Odyssey* is not the *Thousand and One Nights*, and the cave of the Cyclops is not at all that of Ali Baba and the 40 thieves: no secret formula and no language game are able to open it. Ulysses' lie instead only achieves the opposite result of keeping it closed, because, after all, it only produces the effect of preventing the other giants running to help their fellow creature. In effect, the genuine cunning that allows escape is another, much less immediate cunning, whose consequences were crucial and are still evident today.

First, Ulysses picks out an olivewood stake from among the many cut trees which the cave is full of. Olivewood, as we know, is the most crooked tree in all of the Mediterranean. Then he cuts it at the length of two arms. This measurement implies resorting to the use of what Hermann Weyl (1952; It. trans.: 9–11) notes is the first example of a geometrical conception of symmetry, the double-sided conception, that is, the symmetry between right and left typical of mirror reflection but, even before this, of the human body (§ 0). The vertical line between the head, the torso and the legs represents the plane in relation to which the two arms together constitute the horizontal perpendicular line on which from point P to one end there corresponds a single point P' which lies on the other side with respect to the plane. With this we exit decisively from myth, in which right and left are not at all equivalents but correspond to distinct and irreducible qualities (Hübner 1985; It. trans.: 171–85). Conversely, the symmetrical mechanism contains the concept of identity, the concept of difference and the process which puts the former in relation to the latter, which establishes the terms of their correspondence and equivalence. The abstract nature of this mechanism is expressed by the operation that Ulysses commands his men to execute immediately after having carried out the cutting of the stake: the trimming, the straightening, the transformation of the crooked into the straight, of what is curved, rough and irregular into something smooth, polished, uniform but, first of all, rectilinear. In short: the transformation of a natural form, precisely the one most deviating from the rectilinear, into its opposite, into a straight line, the only form which does not exist in nature (§ 9). In the story, Ulysses is made to choose the most distorted tree exactly in order to prove the contrast between the original form and the derived form, to underline the relevance and exemplary character of the metamorphosis. Since this concerns the difference between nature and culture, the olivewood is the crooked timber of humanity, from which it traces its origins.

Straightening is as such the beginning of technique. But it is also the beginning, for the West, of the application of the symmetrical model to the knowledge of the world. Scouring the Mediterranean pond, recurrences, regularities, analogies soon became clear. Similarly, it becomes clear that only by ordering these around an abstract straight line do they become the subject of foresight, calculable in advance, without, that is, any direct experience of them. This line, called a 'diaphragm', coincided with the longitudinal axis of the basin and divided this latter and the surrounding lands into two sectors, an upper and a lower one: in the first, the temperatures dropped as one moved further north from the imaginary line; in the second, on the contrary these increased depending how far one went south of this line (Aujac 1987b: 152–4; Bianchetti 1997: 73–4). In this way, the model united the three fundamental characteristics on which every symmetrical structure is based: representation, transformation, invariability (Auyang 1995: 32–8). Through its application, what had until that point been a series of empirical observations was transformed into the most important idea in understanding the distribution of humanity: the idea of climate.

51. *Men (Women) and Climate*

If, therefore, technology (at least the kind whose model founds modernity and which persists today) begins with the straightening of the stake, science begins with the abstract straight line, unthinkable without this operation. It was only at the end of the nineteenth century that we began to record long-term meteorological observations numerous enough as to allow the construction of an approximate picture of variations in the Earth's climate. The first attempts at the classification of climates are therefore very recent, just as our current concept of climate is recent, if understood as the group of experiences related to the weather and atmospheric behaviour which concern a particular region over a certain number of years. What we see today as climate begins to take form towards the middle of the seventeenth century, when adequate tools for the measurement of its components begin to appear (the mercury barometer for pressure, the thermometer for temperature). Before this, climate was something else; it was not a group of variable phenomena but a fixed and stable part of the known world. It was the portion or, rather, the strip of land, exactly between two parallel straight lines, which every half hour distinguished for each band the different length of the longest day, thus the different latitudes.

Climate in fact meant 'inclination' for the ancient Greeks, evidently that of the Earth's axis in relation to the Sun. Long before dismembering the Earth into seven continents (§ 3), Western thought subdivided the *ecumene* into seven climates. Between the second and first centuries before Christ, Posidonius, the most learned of Stoic philosophers, describes the system of the *ecumene*, extending from the estuary of the current Dnieper, the river which flows to the Black Sea, to the Sudan. Posidonius—Strabo will later reproach him for this (*Geographica*, II.3, II.7)—distinguishes five climates and two zones, partitions which have nothing to do with latitude. For the description of the seven climates of the *ecumene*, then, it is necessary to return to the beginning of the thirteenth century, to the treatise on falconry (*De arte venandi cum avibus*) by Emperor Frederick II, which adopts the version of the Arabic al-Idrisi. Above all, we must return to Albert the Great's *Liber de natura locorum*. The first and the seventh climates, that is, the tropical climate and the northern climate, are the extreme climates; the second and the sixth are less extreme; the fourth and the fifth are the most benign and temperate (Glacken 1967: 224–9). The *Encyclopédie* article on climate, written personally by Diderot, realizes the passage from ancient theory to the modern meaning of the word. What does not change is the very existence of a temperate zone in which the majority of the Earth's population is concentrated. This is despite the enormous expansion of the inhabited Earth in the intervening period.

According to Hassinger (1931: 18–36), the original focal areas for life were lined along a strip which included all the places whose average annual temperature was between 15 and 25 degrees (to give an idea: in Milan, which is at the same latitude as the estuary of the Dnieper, it is just over 11 degrees, in Catania almost 17). In antiquity, the main backbone of the population lay along the range of sub-tropical climates in the northern hemisphere, from the Mediterranean to Mesopotamia to India to China, in almost perfect agreement with the theory of climates if we exclude the southern point of the Indian peninsula. This also fully corresponds with Hassinger's thermal zone if the nucleus of Chinese civilization is excluded, moving a little further north. In the last centuries, the centre of gravity of the European population has shifted to the north, attracted by the cooler temperate countries. But still today, despite this move, half of the world population lives between 20 and 40 degrees north, in the middle belt of the ancient climates. If we include cold temperate countries, where today a fifth of all the inhabitants of the Earth live, the northern hemisphere accommodates, between 20 and 60 degrees north, almost two-thirds of all humanity (Ortolani 1992: 25–9). That is to say, if the reduction of the climate from portion of the globe to a group of atmospheric phenomena still allows us to think of the climate within a schema that is in some way symmetrical, this does not at all apply to what was separated from the concept of climate in the modern period: namely, the whole group of men and women who inhabit the Earth.

52. *The Ace Up the Sleeve, the Axis Up the Sleeve*

As Ritter explained, at first sight there is 'no symmetry in the architectonic whole of earthly Totality' (1852: 206–9, 240). But this does not mean that 'the space filled with earthly things' does not correspond to an order or arrangement. The irregular extent and the different distributions of land and water—with the variability of temperatures and the apparently disordered movement of winds which is its result—constitute the fundamental reason for the ubiquity of and the connection between all the components. Under the chaotic appearance of this general interrelation, forms are hidden, Ritter argues, which govern the destiny of humanity and its adventure. At the beginning of the twentieth century, French sociology and historiography invented the fable of determinism in this regard, polemically assigning to Friedrich Ratzel, the last of the *Erdkunder*, the idea that there is a systematic influence of physical forms on the historical process. In reality, the question is much more complex (Farinelli 1980) and, in any case, for Ratzel, much more than for Ritter, this concerned a dialectical relation between natural fact and cultural element—but this is not our current point. Instead, we might underline the Ritterian scope, as it were, of the most recent acquisitions related to the history of humanity in its relations with the environment and the animal and plant realms.

Why did agriculture spread at different rhythms in different continents? Jared Diamond (1997: 135–46) asks this question in a book which has as its subtitle *The Fates of Human Societies*. The response he gives is the following: the difference between American, African and European events depends above all on the different orientations of the corresponding continental axes. The Americas measure 14,000 kilometres from the southern to the northern tip and 4,800 kilometres at the widest point between east and west, with a minimum of 65 kilometres at the Panama Canal. In other terms, the main axis of the American continent, which is longer than it is wide, is oriented between north and south. For Africa, the same is true, even if on a less pronounced level. In the case of Eurasia, the opposite is the case: the distance between the English Channel and the Sea of Japanese, which is almost 12,000 kilometres, exceeds by no small amount the distance between the southern tip of the Indian peninsula and the Arctic Ocean, so that the continental axis appears oriented between east and west. Since all places located at the same latitude have days of equal length and identical seasonal variations in terms of exposure to the Sun, it follows that on the Eurasian landmass there is a continuity and homogeneity of climatic conditions which is unknown elsewhere. Since the genetic programme of plants is directly affected by these conditions, the speed of their dissemination has been very different from one continent to another.

Almost certainly the most ancient area of food production in the world coincides with the current Fertile Crescent, as the French geographers first named the area: the arc of lands which from the southeastern coast of the Mediterranean to Mesopotamia forms a hood over the Syrian desert to the south. Agriculture is documented to have begun here in 800 BCE and it is from here that domesticated spelt, barley, peas and flax were spread at a speed of more than a kilometre per year on average towards Europe and towards the Indus Valley (Ammermann and Cavalli-Sforza 1973). As Diamond notes: at the time of Christ, Middle Eastern cereals grew along the whole 16,000 kilometres from the

Irish to the Japanese coast, along the longest continuous strip of mainland on the globe. To give a contrasting example: in the Common Era, corn reached the eastern North American coast from Mexico at a speed no higher than half a kilometre per year. The comparison is even harsher if we pass from plants to animals. Almost all the species of the Crescent followed the spread of local domestic plants. Instead, none of the domesticated animals of the Andes (the llama, the alpaca and the guinea pig) arrived in nearby central America before the modern period, thanks to the great climatic contrast between the Cordilleras and the warm and rainy tropical plains which separate them from Mexican reliefs. This is an extreme case of the generally greater difficulty of the longitudinal transmission of animals and plants compared to latitudinal transmission (Gourou 1982: 130–3)—a rule which Ritter only caught a glimpse of (1852: 183–205).

53. *The Cartography and Geography of Genes*

It is a rule, however, which seems valid only up to a certain point in the case of genetic transmission among humans, at least according to the available cartography. Genetic resemblance between one population and another decrease regularly with the growth in distance (§ 35), and this occurs because all populations exchange individuals with their neighbours, through migration. Over time, this phenomenon causes a strong correlation between genetic distance and geographic distance (Malécot 1969). Things are complicated by the fact that gene frequencies are constantly altered by two other factors, natural selection and genetic drift, which is the fluctuation from one generation to the next owing to the random sampling of sperm and eggs. We are talking of interrelations which are very complex, whose outcome, basing ourselves on the relevant maps available today (Cavalli-Sforza et al. 1994; It. trans.: tables I–VII), is nevertheless somewhat expressive even if, given the current state of knowledge, little more than indicative. It must be specified that these are concise maps obtained by means of the representation of linear combinations of gene frequencies in a given population. In this way, the maps are particularly suited to showing the gradients originating in migrations, because these gradients, which, unlike selection, act on all genes in the same way, themselves have a linear effect on frequencies (Menozzi et al. 1978).

In explaining migrations, geneticists make recourse to the concept of 'expansion' and to the model of the 'wave of progress' (Cavalli-Sforza et al. 1994; It. trans.: 205–7). With the first, they indicate the intensification of the occupation of a region or the occupation of new regions on the part of a given population, often following the stimulus of cultural developments which change a relationship to the environment. The entire history of the population of the Earth is understood as a process punctuated by these expansions and waves, starting from Africa and western Asia. From these parts of the world, as far as we know today, the *Homo sapiens sapiens* then spread, that is, expanded, towards eastern Asia, Europe, America and Australia, in accordance with rates of progress dependent on the local density of population and the distance from the starting point.

Only in Africa were the three main ethnic groups clearly organized according to horizontal genetic gradients, organized roughly parallel to the Equator: Caucasian Africans from the Middle East in the north; the black African ethnic group in the centre; the Saan and Khoikhoi, partly related to Asians, in the south. In other continents, conversely, the difference between northern and southern populations seems more or less secondary with respect to the prevalence of vertical gradients which impact more strongly on the distribution of members. This is the case of Asia, within which the main distinction concerns the Caucasian group to the west and the Mongolian group to the east. It is the case of North America, split between native North Americans and Na-Dene speaking populations, and South America, within which the Andes guide the main migratory movement along its axis. It is even more clearly the case of Australia, although our picture particularly suffers from the fragmentary and scarce data. Perhaps, also because of the greater amount of information available, it is Europe that instead displays a more detailed genetic make-up, not definable using boundaries organized in a single direction, and dominated by two major migratory flows: that of Neolithic farmers coming from the Near East, and moving north-west towards the Balkans and west across the Mediterranean; and that of the nomads of the steppes, also moving from the east towards the west.

Limiting ourselves to Europe alone, we can understand that genetic migrations are much more complex than plant and animal migrations, if only for the fact that much fewer fall within the horizontal arrangement of different climatic bands (§ 51). The arrangement along the north–south lines which distinguish a gene gradient from another prove quite the contrary. But a set of genes is neither a man nor a woman—it is not an individual (§ 36).

54. *Isthmuses*

Every living organism is in fact the unique consequence of a history which is in turn the product of the determination and interaction of internal and external forces, often given the collective name 'environment' (Lewontin 1992: 34). Two-thirds of European genetic barriers coincide with geographic boundaries (seas or mountains) as well as linguistic boundaries (§ 35). But charts of genetic distribution that we have seen up to now also fully confirm the role of a boundary on the Earth which is less clear and obvious, appearing less peremptory because it does not correspond to any visible material barrier: the isthmus. As the dictionary explains, an isthmus is a strip of land which puts two solid land extensions of considerable size in communication. It is a large natural bridge laid between continents, as Ratzel explained (1899; It. trans.: 382–84, 209–10). For Ratzel, the isthmus, better than any other form, lent itself to the demonstration of the concept of the 'geographic position' of any individual piece of land: that is, the result of its size and shape, but mostly its capacity to spread (and receive) influences and exchanges with other parts of the world. Between the end of the nineteenth century and the beginning of the twentieth, the isthmuses of Suez, Corinth and Panama, the thinnest and flattest of Euro-African and American mediterraneans, were transformed into straits, and drawn thus into the systems of maritime communication. Until then they had functioned, like others, as zones of conflict and transition between different peoples and cultures, and the genetic charts related to today clearly show this: Central America, for example, exhibits a very complicated mosaic, normal in an area crossed many times by different ethnic groups, but South America shows the predominance of genetic components that are absolutely absent in North America (Cavalli-Sforza et al. 1994, It. trans.: table V).

However, the investigation into the distribution of genes confirms still more clearly the separating function of large continental isthmuses. The clearest cases are the Ponto-Baltic and Stettin–Trieste isthmuses, the two large ideal lines which link respectively the estuary of the Vistula with that of the Dniester and the estuary of the Oder with the Gulf of Trieste. To the east of this bottleneck, there is the thick and solid body of Europe surrounded by closed seas, which merges into the expansive Asian landmass; to the west, there remains instead a shoestring, gradually becoming thinner, the jagged and frayed islands of Europe among the open seas of the Atlantic and the Mediterranean. It is between these two isthmuses that the transitions which allow one to distinguish continental Europe from maritime Europe are located and take place, concerning, above all, climate, vegetation and history. Not only does the 750-millimetre line of rain per year fall between them but also the progression of the winter isotherm to 2 degrees below zero. That is to say, as one moves away from them towards the east, the temperature in winter and the precipitation regularly decrease. Conversely, the thermal gradient suddenly increases: the difference between the average temperature of the coldest month and that of the hottest. This results, on respective sides of the isthmus strip, in a Europe of a mild and humid climate and a Europe of a harsh and dry climate, at the point where the vine and the beech tree disappear and the tundra and the steppes appear. In the absence of consistent physical obstacles, after the fall of the Roman Empire, no invasion coming from the east (Huns and Avars, Bulgarians and Hungarians, Turks and Mongolians) crossed the western isthmus between the Baltic and the Adriatic with all its forces (Dainelli 1933).

This is duly documented by the geography of human genes. The component that derives from pastoral nomads and the peoples of the steppes stops precisely at this strait, beyond which the domination of Germanic and Mediterranean populations stands out (Cavalli-Sforza et al. 1994; It. trans.: table IV). The Ponto-Baltic isthmus and the Stettin–Trieste isthmus function thus also as genetic thresholds, a sign of the impossibility of separating the specificities of human biological make-up from the specificities of the features of the land, that is, the features on the surface of the Earth. And a sign, too, of the fact that long before the transformation of atoms into bits (§ 25), the most powerful borders were often those which could not be seen.

55. *The Woman (Man) Is Flighty*

According to an old saying from French human geography, man is the most mobile of all living beings, and it is clear that here 'man' stands for all men and women. What has been written about the paralysis of the geographic subject remains true, an expression which covers both who produces geographic analysis and who the object of the geographic analysis itself is (§§ 37–38, 40–44). But it is also true that the mobility of human beings is an extremely important part of the functioning of the world, which not even geography has been able to ignore altogether, albeit only assigning the phenomenon a secondary importance in relation to the main topic of discussion: the forms of settlement, meaning not the process but the result of the process itself, the settlements (Farinelli 1981b), understood as the mark of humanity's process of putting down roots. It is not a coincidence that the most thoughtful and systematic of Vidal de la Blache's descendants, Max Sorre, felt the need for a premise on the duality of mobility and stasis, in devoting what remains the most complex and detailed geographic analysis to the topic of human migration. Sorre further explains how the sole reality is movement, and how permanent residence is nothing other than an illusion derived from the slow pace of movement itself. The statement is at odds with the teaching of the master, for whom cartographic representation was the foundation of stability and the certainty of knowledge (§ 40). For Sorre, every image, even the most seemingly clear and homogenous, is a composite and fleeting image, whose stability is always relative and never absolute (1955: 14). We may think that this means the end— implicit and thus silent—of the human geography already described as classic in France (Claval 1984; It. trans.: 67–114). This happens just because the object of geography returns to movement, and with it the mind of the geographer, which is no longer blocked by adhesion to the cartographic model of the world in which all the elements that make up the world appear without life and thus immobile.

Sorre distinguishes primarily between organized group migrations which happen by their own means; individual migrations for work (or those happening in unorganized groups); and periodic movements unrelated to work. The first can be definitive, can have an unlimited range and can or cannot involve the foundation of new settlements: in this case, we distinguish between (a) warrior migrations such as those of the Angles and Saxons towards the British Isles around the middle of the first millennium CE; (b) migrations of hunters, shepherds or farmers who had exhausted the capacities of the terrain, such as most early medieval Barbarian invasions; (c) advanced forms such as those connected to the colonization of America, Asia and Africa in the modern period by European states. Otherwise, organized group migrations can have rhythms, within a specific scope, based around a settlement that is more or less permanent: in this case, they include (a) the journeys of shepherd nomads of the steppes and deserts, such as those of the tribes which still today move towards the Hindu Kush from the Indus basin, returning in autumn; (b) forms related to types of gathering, hunting or fishing, or to itinerant agricultural techniques which are still widespread in southern Asia, based on spontaneous fires destroying vegetation, as well as on continuous movement; (c) semi-nomadism of agricultural work and shepherd farming in mountain regions, such as transhumance, which was practised in Italy until the middle of last century by the shepherds of the central Apennines, who wintered with their flocks on the plains of Puglia and Lazio (§ 23).

Unorganized or individual migrations for work are (but here we must already say *were*) seasonal, such as those in past centuries in Mediterranean and European mountain regions, whose inhabitants moved in winter to work on the plains. Sometimes these movements took place on a large scale, even between one continent and another, often triggered by reminders of cultural identity: this was the case for Emigración Golondrina, the workers between the nineteenth and twentieth centuries who every year after the harvest in Italy went to carry out farming work in Argentina, playing on the climatic contrast between the two hemispheres. It will be useful to reflect on this type of movement briefly.

56. *Life and Kinds of Life*

According to one estimate, on the eve of the First World War, the *golondrina*s, that is, the 'swallows', numbered around 100,000 (Bade 2000; It. trans.: 175). The importance which this transatlantic migration is afforded in manuals (George 1959; It. trans.: 227) is in general the effect of the translation into a kind of folklore of a process which in reality puts in crisis, first of all, the logic of the geographic classification of population movements. One must pay attention: all cyclical and seasonal movements actually obey, except this one, Euclidean criteria (§§ 4, 24) of continuity and homogeneity (of the field within which the movement takes place) and isotropy, which is to say, the reference to a single centre. They are, thus, conceived as though they occurred not in the real world but on a map, on the material support of cartographic representation, which they end up assuming the features of. Geography, it is said, is until now supposed, with some caveats, to describe the world precisely in order that this all occurs (§§ 1–2). This doesn't happen directly, however, but, rather, through the mediation, in this case, of the most powerful and controversial of concepts elaborated by Vidal de la Blache (1911) and his school: the concept of kind of life, the set of practices, techniques and mental models through which a human group survives within a specific physical environment. Set out at the beginning of the twentieth century, this concept did not match up to the analysis of societies based on division of labour, and on the double, connected distinction between social and professional differentiation. This despite the revision, thanks to Sorre, of the original formulation, based on the bijective relationship between environment and kind of life. For Sorre, on the contrary, in every environment there are as many kinds of life as professions (1948; 1952: 11–37; 1957: 197–9): a solution which precisely overlooks the articulation of society, because, for example, doctors and engineers, who have different professions and rhythms of life, belong to the same social class, which instead the young handyman studying engineering is not part of (Derruau 1961: 107–13). With this, the geographers leave the scene and the sociologists firmly enter (Le Lannou 1949: 147–51; George 1966).

It is a shame that the sudden exit of geographers has impeded, in this respect, the understanding of the hidden but decisive function of the idea of kind of life: to transfer the continuity, homogeneity and isotropy of a given environment, made up of fixed and static elements or elements that change their form and position very slowly, to the social composition of those who live within it and move within and outside of it. For this very reason, the migration of 'swallows', disappearing in the interwar period, put this schema into conflict, and as such was taken up primarily as a folkloric phenomenon. This is because crossing the ocean negated the continuity of the medium (land and sea, not simply mountain and plains as in many other cases) within which movement happened; because involving a journey from one hemisphere to another affirmed the existence of an environmental homogeneity in contrast to Herodotus' law according to which the farther a place is from another, the more different it is (§ 25); and because subdividing the year into two periods and two residences for migrants, which were almost equivalent in length and characteristics, put into question isotropy in favour of a sort of bipolar system, a dual orientation.

In reality, if for the collective subject, endowed with Euclidean properties insofar as it possesses a kind of life, we substitute the individual, the overwhelming majority of population movements today on the whole appear discontinuous, non-homogenous, an-isotropic. These are those which, more or less stable and final, Sorre kept for the end of his classification, and which, with the exception of periodic flows for reasons of tourism or religion, concern forced migrations (routes) and all the other forms of displacement made inevitable by economic necessity or economic opportunity or for political reasons: a phenomenon which does not stop growing in frequency and intensity, sparing no region of the globe, and upon which the lives of a growing number of people depend.

57. *Hamlet's Mill*

The finest book of geography of the second half of the twentieth century was written by a historian of science and an anthropologist, Giorgio de Santillana and Hertha von Dechend (1969). The book—*Hamlet's Mill*—takes its title from the mill which, according to the ancient Nordic saga, ground rocks and the limbs of living beings into sand. The mill was moved over the course of the year by the oblique position of the eclipse, by the incline of the Earth's axis in relation to the trajectory of the Sun

(§ 51). This means, in short, the Earth itself, which, in our day, tirelessly shatters the unity of political, social and cultural formations and scatters the bodies of women and men from one place to another—who are, in this way, forced to rediscover in living flesh, violently, the illusory character of every vision of the world based on unity rather than multiplicity, or on the stasis of things rather than on relationships born out of flux: the antinomies lying behind every mythical story and, therefore, all Western knowledge.

Some facts. It was calculated that in the 1930s, no less than 600,000 Jews left Germany, fleeing from National Socialism and finding refuge in more than 80 states (Bade 2000; It. trans.: 306). The Second World War produced around 60 million refugees, exiles and deported people—almost a tenth of the entire European population, if we include the European part of Russia (Fischer 1987: 44). The figure is equal to the number of European citizens who crossed the Atlantic to North America between the first quarter of the nineteenth century and the first quarter of the twentieth (Nugent 1992: 78). The end of the last world conflict also signalled the end of the colonial period and the start of the process of decolonization, which attracted around 7 million people of European origin back to Europe (Emmer 1993: 309), activating at the same time the recall function of this continent. In the course of the 1960s, Europe became an area of immigration, from having been an area of emigration, in the sense that for the first time its net migration was positive. In the meantime, more than 15 million Europeans, starting from the beginning of the 1950s, moved in search of work within Europe (Bade 2000; It. trans.: 342), and the fall of the Berlin Wall revived the tendency: it is estimated, perhaps excessively (Tonizzi 1999: 129), that the disintegration of the Soviet Union made eastern Europe grow from 30 to 50 million people in the first half of the 1990s, that is, from a tenth to a sixth of the whole population (Santel 1995: 117).

In other continents, too, thousands of millions of people moved over the last century, or were made to move in the course of enormous migrations. The extreme case today, though in different proportion, is that of Palestinians, who amount to around 4.5 million people subdivided into just over 3.5 million refugees and just under 1 million migrants: the first in Jordan, Syria and Lebanon, after Gaza and the West Bank; the second in Saudi Arabia and in the other countries of the Persian Gulf, in the American continent, especially the United States, in North Africa and in Europe (Gresh and Rekacewicz 2000). At the other extreme, and at the opposite end of the Arabian Peninsula, more than a third of the 600,000 inhabitants of Bahrain has moved there—coming, for example, from the closest states but mostly from India, attracted by the petrol boom of the 1970s and 80s (Stork 1996). In general, the immigration of Africans and Asians to European cities and countryside, fraught with political and social tensions, represents a tiny drop in the gigantic ocean of transfers of labour within poor countries: only the 20th part, according to the widest assessment, in relation to the whole second half of the last century (Bade 2000; It. trans.: 328). After which, in Asia alone, more than 35 million people, mostly women, have moved from one side of the continent to another for work, a tendency that seems to be constantly increasing (Lim and Oishi 1996; Morice 1997). Evidently, the wheels of Hamlet's mill turn increasingly quickly. Just as for the Ancients, the existence of this mill depended on the discrepancy of the eclipse with the ideal plan of the celestial equator, so today it is fed by the difference between the movements of humanity and an ideal plan of an entirely different kind: that of the global state order—also geometrical and also unaccepted.

58. *A Half-Turn*

Ulysses too turns, as is explained in the first verse of the Odyssey. Indeed he makes many turns: the term *polutropos*, which is the first adjective reserved for him, means precisely that he has turned a lot, that he has made many turns, even if as children they told us that it meant, instead, a many-sided genius. Before being scams, tricks and brainwaves, Ulysses' turns are physical and material, not mental, implying the movement of the whole body and not only that of neuronal circuits. The most important turn among all those he makes is the half-turn underneath the stomach of the creature which serves as a hiding place and which he attaches himself to in order to save himself and to finally escape from Polyphemus' cave.

This turn makes Ulysses into the first subject: literally, because subject comes from *sub-iectum*, which means that which is underneath. Frankfurt School philosophers speak, in this respect, of the 'mimesis of death' (Horkheimer and Adorno 1947; It. trans.: 66) and refer to the immobility of the hero, paralysed by the fear of being discovered if swept up by the hands of the giant who sifts through all its flock in search of enemies. But it is also true that the subject is mobile, because it is transported by the beast. It thus enjoys both conditions of stability and movement, and the first is functional and subordinate with respect to the second: not forgetting that Ulysses' goal (and that of his comrades) is flight and it is only in order to make this succeed do they hide themselves and pretend to be corpses. Philosophers have been clear for some time that the subject moves. Their problem is how it succeeds, in moving, to satisfy the role of founding what exists, to turn over 'the inevitable instance of founding, necessary to give sense to things' (Natoli 1996: *vii*). In this respect, too, Descartes's error is clear: he reduced the subject to thought and he threw away the body. We tend to forget, however, that a long time before Descartes, someone had already eliminated all the rest, which in the *Odyssey* is instead described with extreme precision: the cave, the giant, the beast, in short, all of history, to finish with the nature of movement on which depends not the existence but the survival of the subject.

What in any case is certain is that the movement in question, that of the creature, is not limited to conducting from within the cave to the outside, into the open air, but transfers the subject from place to space (§ 3). It is precisely in this sense that nothing is any longer as before, for Ulysses and his comrades relinquished to the giant's fury (§ 0). A place, it has been claimed, is a 'field of attention', whose force depends on the emotional investment of those who use it. Different from a monument, a place cannot be known from the outside, but only from the inside, and it is tightly connected to our identity which is something definable only in relation to others. For this very reason, every place is a small world, in the sense of something which depends on a set of relations between human beings (Tuan 1974: 233–46). The first place which the West conserves the memory of is Polyphemus' cave, which satisfies all the conditions of the definition previously mentioned, with the sole variation that in this case the relations do not only concern Ulysses and his comrades but extend to those between men and all the other forms of life. At the same time, it has been noted, characterizing place as home, as a stable and immutable environment to return to is per se a masculine attitude (Massey and Jess 1995; It. trans.: 54). It would be a mistake to think, in this respect, of Ithaca, which Ulysses never references in the *Odyssey* with regret or nostalgia. Before any distinction of gender, in a world where even the difference between what is animate and what is inanimate is problematic, as in the case of Polyphemus, places are locations which are anything but peaceful—they are themselves seats of conflict and change: as the example of Ithaca itself confirms, on Ulysses' return. Ulysses' journey between the land of the giants and Ithaca unfolds between two places only because in leaving the cave, and in order to save himself, he invents a new model of the world which transforms all the parts of the world which escape him—space—into places, that is, into the opposite of the self.

59. *The Line of Flight*

To exit from the monster's den does not yet equal salvation. This will truly be reached only when Ulysses reconnects with the rest of the fleet which is waiting on the beach in front of the land of the Cyclops. It is precisely in the stretch of sea which separates the two shores where space makes its first full appearance. Ulysses' trick negates, inside the cave, the first rule of the world, that in which order depends on the existence of levels and on the coincidence between the role and the position of things. To hide oneself underneath the stomach of beasts rather than to ride them equals, in fact, before all else, treating the two levels of the animal body, the upper and the lower, as simple dimensions: no hierarchical relationship is recognized—which instead is implicitly negated and subverted—but, rather, only specific, equivalent dimensions. What is vital is not their relation but their surface. At the same time, any possibility of inferring the function (of human beings) from their position is put into crisis. This, instead, is normal in any hierarchically ordered structure, because the relation is immediately visible. One thinks of the seat of any society, and of an office occupied by its employees and managers, who is placed higher or lower in the office according to their importance, or according to the greater cost of the top levels in relation to the lower levels of any residential structure (this, incidentally, has only been the case since lifts were invented, that is, since buildings were definitively transformed into spatial mechanisms). His being invisible is Ulysses' problem, whose genius, as is normal in the Greek world (Detienne and Vernant 1974: 10), is applied to what is mobile and escapes rigorous reasoning, exact calculation and precise measurement: to all which, in short, is exactly the opposite of what space means (§ 3). In order for space to appear, further tests are required, pointed in the same direction straight ahead: the direction of flight.

Returning into the open, Ulysses must first re-establish his authority over his comrades, which had been put into crisis by the tremendous trouble he had thrown them into. For this reason, too, he violently addresses himself to the already blind Polyphemus who has remained on the edge of the coast while ordering his men to hurry up and put as much distance as possible between the hull of their ship and the giant. Polyphemus' reaction to Ulysses' invective is terrifying: he turns towards the shouting and throws a boulder which falls at the ship's bow, and the resulting wave leads the boat back to its starting point. Going back out to sea, Ulysses does not hold himself back and shouts his true name at the Cyclops, who again throws a boulder towards the voice; but this time the projectile falls astern of the vessel, no longer in front of it but behind it, so that the wave produced pushes the boat forwards, depositing it in one swoop onto the beach where all the other Greeks wait. Why does the first shout have an inauspicious effect, involving a return to the gravest danger? And why, instead, is the effect of the second, even if indirect, that of directly and instantaneously bringing them to definitive salvation?

The first time, Ulysses shouts when he claims to be distant but just in earshot of the giant, that is, when he fears that to keep waiting a few more moments could compromise the possibility of being heard, could impede his invective reaching its destination. In other terms, his is not a calculation but an estimate, a valuation which concerns the relation between two physical performances, between two corporal functions: those of his own throat and the ear of the enemy. But the estimation does not work and the result is disastrous. The second time, instead, it is a true calculation, which works past all hope: Ulysses, the text says, waits to have crossed twice the distance than before. How the hero could have been able to achieve this new measurement concretely is not an easy thing to establish. What is crucial, however, is that salvation is the result of abstraction from all material referents, from any relation of a physical nature. To double the first distance, which is merely a mental issue, equals the passage from one to two on the Pythagorean table (§ 43), implying the existence of a standard interval such as that which exists between a natural number and the following number. It thus implies the recourse to a formidable general equivalent: space itself (§ 85).

PART THREE

The City, the Map, Space

60. *The Index: The Road and the House*

'Ships slide through the water, the cleft waves roll together, and all the trace of the passage is blotted out. But land preserves traces of the routes travelled by mankind. The road is branded on the soil. It sows seeds of life—houses, hamlets, villages and towns.' Thus Vidal de la Blache (1926: 370) imposes his settlement geography on the indexical relation between means of communication and houses. This is even stronger and more systematic than the relation Peirce (§ 14) individuated between the smoke and the fire: there can in some cases be, strictly speaking, smoke without fire, but there cannot be a house without a means of communication, however slight and small it may be. This is a relation based not only on space and time but also on a causal link, as Vidal, with the reference to the generative nature of the road, first emphasizes. In reality, the link between house and road is so intimate that we must pose the question: On what basis and to what extent can we separate one from the other, and what sense does this have? It is true that the development of settlement geography is based, instead, on the divorce between the facts of settlement and phenomena related to circulation (Farinelli 1981b: 12–32). So that claiming the organic character of the relation means putting all of twentieth-century human geography into discussion, in the same way as Jane Jacobs has criticized urban thought, claiming the indissoluble nature of the relation.

When we think of a city, Jacobs maintains, the first thing that comes into mind are its streets, and when we think of the apparent disorder of old cities we must instead realize the complex social order guaranteed by the close mixing of different urban uses along pavements: an order that consists in the functioning of the most varied activities, one next to another, each possessing its own rhythms and schedules, and which together translate into direct and continuous supervision of the streets themselves. On the contrary, the whole of modern urbanism is built on the opposite principle, on an order based on segregation, that is, on the distinction and separation of a certain number of elementary uses of the citizen soil, which are each assigned to an independent and isolated location. This occurs because it is not the street but the block which is understood as the fundamental unit of architecture. The result is that, although animated by the best intentions, architects and urban planners believe blindly in what they were taught at school about how cities should function, so that they never manage to understand how these actually function in the concrete everyday. Quite contrary to this, the real order, which is exhibited every day by every street in the old city, is 'fact of movement and change, is life' (Jacobs 1961; It. trans.: 27, 46, 23, 17, 18, 7).

Jacobs' critique stops here, and arguably it can be compared to Mandelbrot's critique of (§§ 10, 34) on the prevailing of the model over reality. His last observation also takes us back to Ritter's old polemic, to his contrast between the life of the world and its cadaverous, that is, cartographic, representation (§ 8). It is precisely this call to uncover in the logic of the map the schema which even Jacobs unconsciously reacts to: the same schema for which settlement geography functions as a loyal protocol. Let's see how.

Ratzel's anthropogeography (1899; It. trans.: 85, 95) already had the ambition and the consciousness of being a philosophy of history, and its goal was the understanding of the 'mutual influences which are exercised between the people and the territory and between the territory and the State'. Thus, it was necessary to 'turn our gaze towards the past, to seek there the causes that the present does not show us' (Ratzel 1899; It. trans.: 85, 95). Conversely, for Otto Schlüter geography is no longer knowledge relative to a series of problematic relations, but for the first time becomes a 'science of the object' (1906: 10–11, 26–7; 1919: 18). According to

Schlüter there exist only two types of science: the first considers phenomena according to their temporal becoming, and these are historical sciences; the second considers them according to their being, thus as simple objects, which are the sole things that we can distinguish, and it is this type of thinking that geography belongs to. As though it were easy to establish what an object is.

61. *Once There Was a Beautiful House*

In reality, for geographers of the last century it was actually very simple: the object was what resulted from the cartographic form, or, better, the topographic form (§ 41), of things. For Humboldt (1845: 171) form (*Gestalt*) attests to the mode of formation of the object of geographic investigation, 'this is its history', and thus the visible expression of a process that, on pain of failing to be understood, had to be reconstructed. What Schlüter (1906: 22, 24) instead refuses to tolerate is precisely description that, as in description by historians, begins from the *Gestalt* which he makes correspond to 'individual vision'. Oscar Peschel (1876: 3–5, 68) had already substituted the term 'geographical homology' for this as a starting point, meaning the examination of similarities between the same structures as they are represented by the cartographer: he had in this way reduced, in the field of geomorphology, geographic analysis to cartographic description. Schlüter (1952: 89; 1899: 67) transfers the master's lesson to human geography, whose goal is reduced to the study of the 'simple form (*Form*) visible' on the topographic chart. It will be precisely this form, cartographically determined, which will govern the entire settlement geography, which, together with population geography, will divide, for the whole of the twentieth century, the field defined in the eighteenth century by the problematic relations investigated by the *Erdkunde*. This is despite the indexical link suggested by Vidal. Proof of this is the classification of rural houses developed by his student and successor at the head of French human geography, Albert Demangeon.

This is an as ever-convincing example, because Demangeon's starting point is the very recognition of the historical nature of the object in question, product of a long evolution, and work tool which epitomizes the experiences of many generations of farmers. He suggested investigating not according to materials of construction or exterior forms but according to internal structure, according to the relations which it establishes between men, animals and things, which is to say, according to their agricultural function (Demangeon 1942: 262, 263). Yet the typology thus set up mortifies somewhat the correctness of the intention. Demangeon distinguishes two principal types of rural houses: (a) the block or unit house, also called the 'global house' because it is composed of a building where everything is found under the same roof; (b) the courtyard house, in the cases where two or more buildings are found grouped around a cleared area, the courtyard. The first is the habitation of poor farmers, who live from working a tiny piece of land. It must be noted that at least until the middle of the twentieth century, more than half of the population of the globe lived in houses of this type, often composed only of clay and wood (Gourou 1973: 211; Ortolani 1984: 53, no. 6). According to shape, four principal varieties can be identified: (a) the basic house, in which stables and lodgings are next to one another; (b) the block house with transversal elements, that is, juxtaposed elements, on whose facade the three essential elements are lined up—lodgings, stables and barn; (c) the block house with longitudinal elements, elongated in depth and not in facade; (d) the house with overlapping elements—lodgings above, the rest below. Instead, the courtyard house, an expression of more complex and structured economic and social relationships, presents only two varieties: (a) the closed courtyard, in which the buildings are touching and completely surround the courtyard on their inside; (b) the open courtyard, in which the buildings are not touching, and the enclosure of the uncovered area onto which these look out remains partial (Demangeon 1942: 230–5).

From this classification there follows, in fact, an evident paradox, a product specifically of the structural ignorance of any indexical relation:

the block house does not possess a courtyard. Naturally, it is never like this: precisely because Demangeon's premises apply, there cannot exist a farmer's house, however simple and small it is, without a clearing in front of it, which is absolutely indispensable for the evolution of daily life and where often, in fact, the largest amount of time is spent (Farinelli 1977: 82). How, despite these presuppositions, can settlement geography itself amputate the object of its functionally most important part?

62. *Cartographic Rhetoric*

However strange it may seem, the answer is exemplarily illustrated by the evolution of Italian geography. It is a response which involves the whole nature of geographic knowledge. More than half a century before quantitative geography (§ 42), at the beginning of the twentieth century, Olinto Marinelli recognized in cartography 'a true instrument of thought', capable of the 'materialization of complicated relationships', that is, of the 'mechanical simplification of ideas' (1902: 234–5). Consequently, continued Marinelli: 'we can unite all the general ideas related to our discipline around the map, and we can unite all the special ideas around the topographic chart' (ibid.), ideas related to individual regions of the earth. This is a scientific programme very reminiscent of Peschel's, adding to it a greater level of consciousness about the control established by the mechanism of the scale on the level of conceptualization (on the idea we make) of the mapped phenomenon: something which, after Marinelli, geography in general, not only Italian geography, will completely forget—only to rediscover it traumatically at the start of the last quarter of the twentieth century (Lacoste 1975; It. trans.: 185).

This rediscovery significantly follows close behind settlement geography's rejection of all typology based on planimetric and topographic considerations such as those of Demangeon. The argument in question is precisely the form of the closed-courtyard residence which appears throughout Italy from the Po Valley to the islands: if the form is largely the same, it is shown, this is an expression of socioeconomic realities and historical processes which are profoundly different if not contrasting. From this standpoint, we can delineate the polarity between the large buildings of the capitalist company, based on monoculture and the employment of salaried labour, and the small agricultural business which is located in single-business residences or residences with multiple businesses that are all direct cultivators, adapted primarily to polyculture. In some areas, as in the Lombardy-Piedmont plains, the distribution of these types reveals an almost perfect congruence of two different natural environments: the lower irrigated zone in the case of the first, the high dry zone and the corresponding strip of hills in the case of the second. But the coincidence between physiographic region and form of habitation is not direct, and would not exist without a diversity of agrarian structure which the residence is the vessel for (Pecora 1970: 219–20, 234–5, 237).

In fact, the refusal of simple formal or planimetric analysis removes all the ontological power from the cartographic dispositive, its capacity to determine the nature of the examined object. In the case of the typology of settlements, this does not so much concern the transformation of the product of a historical and social process into a physical and natural phenomenon (§ 41), but the reduction of the object, according to the mechanical procedure of the scale, into a synecdoche or metonym of itself. The residence, that is, the building, stands in for the business which is founded on it, and which it is thus part of (Farinelli 1976: 634–5, table 104). 'Geographical homology', to take up Peschel's original expression, is based therefore on rhetoric and its figures. Precisely because the effect (the group of buildings, the house) substitutes the cause (the business, the set of productive factors and relations of production), settlement geography, which is a European invention, ceases to exist completely at the beginning of the 1960s. These are the exact years when in the anglophone world, quantitative geography (§ 42) also enters into crisis, and essentially for the same reason: it is realized that one cannot infer the process from the form. Even in the realm of mathematic and geometric symbols, much more abstract than topographic ones, contrasting processes produce identical forms, identical profiles. This will lead the most heretical geographers to pose the problem of the meaning of similar representations, transferring their reflections on the description of physical phenomena towards investigations on the deep structure of argument (Olsson 1974; It. trans.: 52–5; 1980).

It remains to clarify how cartographic rhetoric acts concretely, and how topographic representation manages to transform cause into effect. It remains, in short, to reveal the deadly gimmick of the 'graphic prejudice' (Febvre 1922; It. trans.: 68).

63. *The 'Graphic Prejudice'*

For Lucien Febvre, this expression serves to characterize the attitude of those who content themselves with a formal resemblance in order to compare, on the basis of a cartographic outline, things which have nothing in common from the genetic point of view. It serves essentially to criticize the geographical homology which, like every homology, derives the proof of the evolutionary relation between two organisms from the similarity of their forms (Bateson 1979; It. trans.: 302): precisely the kind of relation, as in the case of the courtyard residence, that Pecora shows does not exist. But before being in the mind of those who look at the map, the graphic prejudice is nestled inside representation itself, and the cartographic profiling of things is already a reflection of it.

Up to this point, the isolated rural residence and the particular kind of building (that is, the very reduced nucleus of habitations) which corresponds to the large capitalist courtyard have been taken as examples. Let's now take the third of the forms of settlement which are distinguished in Vidal's quotation (§ 60) from which we began: the village. Open any dictionary—it is very difficult to find a definition of village that departs from a group of houses smaller than a city and larger than a building. However, the meaning will be a group of constructions, a complex of buildings, or a built-up group. All this seems normal today; in fact, here we refer not to geographic dictionaries (this would be too easy) but to those we habitually use to know what words mean. However, this should not seem so normal, if in fact we only need to move a small amount in space or time to encounter, for the same term, a meaning completely opposite to the current one. In India, for example, it is held that the village is a piece of land delimited by boundaries, not necessarily possessing habitations but, instead, always possessing a name which signals the existence of rights of usufruct or property rights over the land itself. The village is, then, a land unit and a fiscal unit before being a residential unit (Farinelli 1981b: 10–11). In other words: without terrain on which to place habitations, they could not even exist. Thus, a definition which recognizes the prevalence of the former over the latter seems logical, that is, the priority of the base of the relations of production (land) in relation to the product of work (houses, which in the Indian countryside are also often made of clay). It only takes a quick skim of the *Oxford English Dictionary* to realize that until a few centuries ago, the village had the same meaning in Western culture which it has today in Eastern culture.

So the question becomes: How, when and why in the modern period did the village start to mean habitation and no longer cultivation? The response is found in the most famous work of polemology of the nineteenth century—*On War* by Carl von Clausewitz, the Prussian general and student of Ritter. Clausewitz explains that the forms of land influence fighting in three ways: as obstacle to sight, as obstacle to passage and as means of cover against the effects of fire (1832; It. trans.: 433). Consequently, it was on the basis of these criteria that between the nineteenth and twentieth centuries, (§ 41) topographic charts proceeded to be made in Europe, selecting according to the criterion of greater or lesser tactical value and transforming all the features of the Earth's surface into an obstacle or advantage. As Marinelli will argue (1902: 236):

the sole motive for the gratitude of geographers towards the army consists in the army's activity as a cartographic survey. In this way, what is fixed, evident and sizeable is systematically privileged, in geography too, against what instead remains mobile, fleeting and small in volume: trees come to be preferred to herbaceous crops which, in addition to being closer to the ground and less voluminous, are often also temporary; the limits of property (and thus of the concept of the farm itself) disappear from view, or are otherwise marked out only by robust fences; etc.

For this, the village, mechanically stripped of every sign referring to its nature as a social formation, is transformed, first on the map and then in our minds, into a simple group of houses, into a collection of constructions.

64. *Between Myth and Archetype: What Is a City?*

The same applies, more so, for the city, whose simplest but most widespread definition still insists on it being an agglomerate of buildings, only larger in extent than a village. But what is a city really?

'A city is a city'. For historians, this obvious truism, coined by Robert S. Lopez (1963: 9) expresses the renunciation, for some time now, of the attempt to make an all-encompassing response to the question: Why does the concept of city change from period to period and from country to country? On the contrary, through epochs and cultures, what does not change at all, or changes only a little, is the level of consciousness of the existence of the city on the part of contemporaries. As such, the confrontation between the development and the function of city-organisms different to one another would be revealed as much more productive than the search for a univocal, and thus supra-historical, formulation of the urban (Berengo 1975). This is what happened in the last quarter of the twentieth century, even though the comparative approach later privileged the character of the city as a political institution, understood as a complex of structures and functions connected to the exercise of power, with few references to its internal mechanisms (Rossi 1987).

This is what historians argue. Geographers, instead, still debate within the contradiction between formal consideration, that is, topographic consideration, and functional consideration of the city, on the basis of the following paradox: if attention is paid to physical dimensions, to the continual expansion of the constructed area, to the greater and even greater size of the urban population, the city today seems to grow and extend outwards; but if, instead, we pay attention to the typical functions of the city (coordinating, leading, controlling the surrounding territory), then we would be led to conclude that the city is rarefying and disappearing, because these functions are becoming more and more concentrated in a limited number of areas of the world (Dematteis 1978: 185).

In the period which the last quotation dates to, according to statistics, the greater part of the population of the globe still lived in the countryside, not in the city. Today, coinciding exactly with the passage from one millennium to another, still according to statistics, the ratio has been reversed, and more than half of humanity (thus more than 3 billion people) lives in cities (whatever this word is taken to mean). In fact, this has rather different meanings from one state to another, in the sense that the minimum number of inhabitants necessary for a location to be classified as urban varies from country to country: 250 in Denmark while 40,000 in South Korea, for example (Ortolani 1984: 32). What does not vary is actually the nature of the criterion which censuses adopt: rigidly quantitative. Statistics, in short, do nothing other than translate into a different language—a numeric language—the topographic idea of the settlement. Statistics reduce the settlement, which is in reality a process, into a series of things and then distinguish those things, according to their size, as rural or urban settlements. The sole difference is that while from the topographic point of view what is important is the form, from the statistical point of view what makes a difference is the format, that is, the very thing that derives from the form.

From here derives the paradox above, since in this way the function, denied entry to the door of definition, re-enters arrogantly and without warning through the window. This is because 'A whole mythology is deposited in our language' (Wittgenstein 2010: 31): a mythology in which, in this respect, scattered houses, farmhouses, villages and cities would be characterized not only by their growing dimensions but also by a growing complexity of the accommodated functions, almost as though there existed a direct relation between the importance of these functions and the size of the building complex. This is a mythological conception of the question if we take myth here to mean a false clarity based on familiarity, which saves the concept from the analysis (Horkheimer and Adorno 1947; It. trans.: 6). However, in the events relating to the origins of cities, myth is never the first version of the facts but always the last (Wheatley 1971: *xv*). In this case too, myth is related to a primordial archetype of which the series of forms of settlements growing in dimension and complexity are the reflection; myth puts these in sequence so that the archetype can order them in a genealogical sense. And vice versa.

65. The 'Invisible Hand' and the Hidden Hand

The archetype referred to is that according to which the emergence of civilization (a word which derives from the Latin *civitas*, city) is the result of a sequence which begins with the activities of hunting and collecting wild fruits, followed by agriculture, and culminates with urban and state formation—a linear, continuous and cumulative sequence which only in the second half of the twentieth century was definitively put into question, and which corresponds to a stadial version of the history of humanity, that is, proceeding through stadia or phases. It should be noted that in this case too, as will be seen, the stadial model, the order based on a series of stadia, implies space, and it would be unthinkable without the advent of the spatial order of the world (§ 3).

The strongest reaction against this archetype, which, for example, even Lewis Mumford (1961) essentially holds to, was that of Jane Jacobs.

Why, she asks, must agriculture necessarily precede the birth of the city? How did anthropologists manage to establish this? Surprised that such a clear and obvious question could be put into the discussion, the called-on anthropologists responded that this was established by economists. In turn interrogated, the economists responded instead to have learnt this from anthropologists and archaeologists. Each, in short, sent the ball back to the other's court. The final guess Jacobs made was that the archetype derives not from the evolutionary schema of Darwin, as we would be led at first glance to hypothesize, but from a previous source: Adam Smith, whose *Inquiry into the Nature and Causes of the Wealth of Nations* appeared in 1776 (Jacobs 1969: 42). The implicit development in the series of settlements, which leads from rural to urban forms, derives in this way not from biological evolution but from historical evolution, not from zoology but from political economy.

This would explain many things, beginning from the rigidly quantitative criterion for the distinction between village and city. What is important to Smith is the technical division of labour, which is something other than the specialization of tasks, and is the cause of the increase in productivity, thus of all conquests made by humanity. But the division of labour, which descends indeed from the natural tendency of men towards exchange, depends in turn on the market, because the market must absorb its product. So Smith considers villages, cities and large cities, above all, as isolated markets, whose extent is measured not by their dimensions but only by the mass of buyers or by the possible sale of any given good (Smith 1776; It. trans.: 19–20). The division of labour, which confers on labour an infinite productive capacity, is in this way a function of reference for demographic size.

It still remains to clarify how and why this vision then became the only possible vision, not only in geography, and has constituted the basis of a general evolutionary theory of urban development. In fact, this theory too, in the same way as the sequence of geographic forms of settlement, was defined according to the logic of the division of labour, in which for every stage of evolution and for every element of the formal series, a specific and distinct task is assigned. For Smith, the final stage of society, the commercial stage, is regulated by the 'invisible hand' which governs the economy and whose activity is described in the first two books of the volume. It is to its involuntary action that we owe the reconciliation between liberty and the desire for the bettering of the individual and the societal order through the mechanism of competition. But as Douglas Hofstadter has shown (1979; It. trans.: 747–48) in relation to Escher's lithograph *Drawing Hands,* in order for an invisible hand to exist, one specific condition is necessary: that the world is reduced to a design which also includes the hand of the author, and which thus negates the concrete existence of the hand, relegated to an 'inviolable level' for the eye. As precisely, it follows, the idea of the city testifies.

66. *'Drawing Hands'*

Escher's lithograph is an illustration of recursivity: a hand draws a hand which in its turn, in a circle, draws the first hand. The cat eats its own tail. This occurs, Hofstadter explains, only so that in this way the image seals reality, occulting the hand of the designer (not drawn but real) which has drawn the image itself. The hand thus hidden is the invisible hand, and the visible hand, that is the drawn one, is the false hand. But all this catches only one aspect of the question, and not even the most important one. The mental cramp which takes hold of us while looking at Escher's image derives from something much deeper, from the necessity of a terrible admission, complementing what has already been said: it is the drawing itself which is the carrier of an intention, a will, which is independent of the intentions or the will (of the hand) of any given drawer. For this reason, Smith's 'invisible hand' proceeds involuntarily, since it follows an intention which is already inscribed, invisibly, in an image. It is precisely the evolution of the meaning of the term 'city' that confuses us as to which intention and which image we speak of.

Let's start off from the Los Angeles City Hall, constructed in the 1920s to equal, despite the dangers of an earthquake, the height of the biblical Tower of Babel, the Babylonian ziggurat (§ 8). A piece of writing on its outer wall is reminiscent of something which Aristotle absolutely agreed with, in terms of having already written it: the city is born to preserve life, existing to make men live well. Until the end of the sixteenth century, people continued to think in the same way. The small tract which sets off modern theoretical reflection on urban nature (Gambi 1973b: 367–71), *Delle cause della grandezza delle città* by Giovanni Botero, begins thus: 'Cities demand a wealth of men, grouped together in order to live happily' (1598: 329). Immediately after he adds something with which, some centuries later, for other motives, Adam Smith will agree without trouble: the size of a city is not given by the extent of its site or by the length of the perimeter of its walls but, instead, by 'the multitude of the inhabitants, and their power'. But in the eighteenth century, the idea of the city experiences an evident and extraordinary overturning: it passes to mean no longer men but things, houses. Men, with Smith, will survive within political economy, only because they are transformed into producers and consumers. The definition of city is transformed into the one that we know now (§ 64), and which is still, too, the codified Enlightenment definition, under the term in the *Encyclopédie*: a 'group of several houses arranged along streets and surrounded by a common element, often walls or ditches'. Immediately this is clarified: 'to define a city more exactly, it is a belt of walls which encloses public quarters, streets, squares and other buildings'.

In looking at the photos of Paris taken by Eugene Atget in the nineteenth century, Walter Benjamin (1955; It. trans.: 29) was astonished, in the twentieth century, to find them to be without men, almost as though the city was empty, without inhabitants. He cautioned against the 'hidden political meaning' of these images, a character of which he did not manage, however, to decipher the nature. To do this, it would have been in fact necessary to begin from the change in the idea of the city developing in the first centuries of modernity: the human absence in the first urban photographs derives from this transition. Why do the first photographs make of every city a ghost city? To what do we owe the forced evacuation of people? Reread the definition in the *Encyclopédie* just given. The verbal, literary definition show the exact calculations in the cartographic definition of the city, in which the mechanism of the scale excludes all human beings: definition starts from the wall, from the most cartographically relevant element, underlining with initial hesitation precisely the obedience to the logic of the graphic prejudice (§ 63). Between Botero and the Enlightenment thinkers, thus, something decisive has occurred. This something is the construction of the first topographic chart, the map of France based on the Paris meridian (§ 9), whose preparation, beginning in 1669, ended exactly in the year of the taking of the Bastille (§ 10). After which nothing in the world was as before. After, that is, the map.

67. *Streets, Roads, Paths*

The map of France, the first great truly scientific map, does not in fact only change the idea of the city or the form of streets (§ 9) but also the form of cities and the idea of the street, a sign of the impossibility of making this distinction, on the map, between form and idea. It also demonstrates a tremendous ontological activity, in addition to productive activity, linked to cartographic work and its outcome.

Let's look again at the *Encyclopédie*, under the term 'path' (*chemin*). We learn that in the middle of the eighteenth century, path, way (*voie*) and road (*route*) were still synonymous. It is therefore determined that 'way' will indicate the mode in which one proceeds, and can be by land or by sea. 'Road' includes instead the set of places which one needs to cross to travel between one location and another: from Paris to Lyons, one passes along the Burgundy or the Nivernais road. Finally, 'path' refers to the selection of land on which one proceeds along one's own route. What is the model for these specifications?

In 1669, Colbert, King Louis XIV's powerful prime minister, had tasked Giandomenico Cassini with the command of the large operations of an astronomical-geodetic survey required for the redefinition of the French territory in geometrical terms (that is, spatial terms). It was his grandson, César-François Cassini, son of Jacques (as de Thury called him to distinguish him from his father and grandfather), who would illustrate before the Academy of Sciences, in 1745, the very close relationship between these operations and the means of communication. His speech brought attention to the fact that, owing to the enormous cost and the scarcity of capable people, only a very small number of maps had been compiled up to that point in a geometric way (*par les voies géometriques*). Consequently, in the comparison between ancient maps and modern maps, numerous errors were shown, since their makers had followed very different roads (*routes*). Some had been satisfied to estimate distances by the time taken to move around. Others had effectively measured, that is, with a measuring tape, the path (*chemin*). Others still, operating with greater scruples, had executed a sort of rudimental triangulation, relying, however, on bases that were too short and these also measured with a string, and, above all, orienting their maps by only using the compass (§ 11). These were such rough procedures that, multiplying the disadvantages in the event of assembling further maps, they led to differences of even up to 30 degrees with respect to the exact course of the meridians. And, as Cassini de Thury reiterates, without exact knowledge of extent, it is difficult to make certain measurements of the boundaries and the position of different places for the purposes of a large number of useful projects for the state and commerce: the construction of new streets, new bridges and embankments, new canals suitable to facilitate the transport of foodstuffs and goods from one region to another, in this way preventing famine and procuring abundance through the increase of traffic and communications. Thus, to establish the general map of France, he proposes the generalization of the same method employed for the 'description' of the Paris meridian (§ 10): to form triangles across the entire breadth of the realm, each linked to the next by means of objects which are successively revealed, each departing from the last (Cassini de Thury 1749).

As can be seen, the distinction in the *Encyclopédie* between ways, roads and paths seems precisely to descend, limb by limb, from Cassini's project for the 'geometrical description of France', from the distinction he introduces between the terms in question. The principal difference consists in the fact that the street, understood by Cassini abstractly as a process, becomes, rather, a virtual process, changing in the *Encyclopédie* into a concrete but only hypothetical succession of locations, precisely as it only appears and can be calculated on the map. If it were not this way, we would not know how to distinguish the road from the path at all, which both for Cassini and the *Encyclopédie* corresponds only to something which is walked upon by foot. That is to say, once again, that without cartographic representation, we would confuse the meanings of words with one another. Or, that the order of language depends on cartographic logic.

68. *Taxi*!

In reality, this makes us think of the fact that Giovanni Botero took his definition of the city from Torquato Tasso (1958: 414–15, 342), who already in 1587 had made the difference between villa and city clear in the following terms: the first is a 'coming together of men and habitations with the things necessary for life'; the second, with the 'things necessary for the good life'. He added that happiness consists, at least in part, in not being constrained, being in a villa, to make recourse to the city 'for the things necessary for living well, as well as living'.

What is without doubt, however, is the almost obsessive role that the figure of the city and its cartographic portrait hold for Tasso. His boast, repeated many times and perhaps the main one he makes, consists in the claim to have been the first to make the city into the centre of a poem: the *Illiad*, he specifies in that regard, does not sing of Troy and the war fought for it but simply of the ire of Achilles (Tasso 1853: 86). When, at the beginning of the nineteenth century, Chateaubriand, *Jerusalem Delivered* in hand, visits the precincts of the sacred city, he is struck by the topographic precision of the description: 'if it had been made at the place itself, it would not be more exact' (1850[1811]: 374–5). Moreover, the main theme of the letters exchanged between Tasso (1853: 86, 133) and his Roman correspondent Luca Scalabrino between 1575 and 1576, the years of the completion of *Jerusalem Delivered*, is only one: the maps of Jerusalem available in Rome and unavailable in Venice or elsewhere. It is not surprising, then, that in the *Discourses on the Art of Poetry* (Tasso 1959: 387), he describes his poem as a work in which things are connected to one another as in a 'little world', as in a city, so that if one part is missing or is moved, all of the rest collapses.

The city that the poem is modelled on is Ferrara. Both Ferrara and *Jerusalem Delivered* are organic creations within which it is impossible to confuse two different and opposed, although contiguous, environments: the pagan world and the Christian world of the poem, the medieval city and the modern city in Ferrara, designed and realized at the end of the fifteenth century by Biagio Rossetti at the behest of Ercole I d'Este. The very structure of this modern city supplies the model for poetic composition, so that both Tasso and Rossetti, the first modern European urban planner, resolve the problem of centrality in the same way: Rossetti with the asymmetrical arrangement of the main square of the new city, departing from the axis of the medieval castle (Zevi 1960: 161), that is, the old centre; Tasso with the asymmetrical arrangement of the narrative whose centre does not correspond to the tenth of the 20 cantos which the poem is composed of, but falls at the thirteenth (Raimondi 1980: 91).

That there is a connection between the architect's and the poet's asymmetry is suggested by an observation which can be verified on the map of Ferrara. Rossetti's addition is subdivided into 20 sections (as many as there are cantos in the poem), along which the new square rises, and it counts from west to east, that is, following the journey of the Crusaders: one can see that the square opens exactly at the thirteenth segment (Farinelli 1985: 78–9). That is to say, the structure of *Jerusalem Delivered* imitates that of the system of new Ferrara in mathematical relations too. Mathematical, and not simply arithmetical, since we are speaking of a model which is also, and above all, geometrical: Ferrara is the first modern city in Europe, as Jacob Burckhardt claimed, because in it, for the first time after the classical era, the orthogonal and rectilinear syntax of streets makes its domineering and disruptive appearance, the only syntax capable of reducing the city to spatial extent (§§ 3–4). That is, in Ferrara, modern urban space is born, next to the ancient medieval settlement, the kind of space we are so used to today that we can no longer perceive it. And which we celebrate every time, in a hurry to make up lost time, we call a taxi, a name which derives precisely from Tasso, whose family had received the general contract for the Imperial Postal Service from Charles V ('Taxis' is still the name of the German branch) —a family which had thus made its profession out of speed, the destruction of places and the transformation of the world into space. This, however, Torquato did not share and perhaps, to the point of dying for it, did not accept.

69. *New Lands*

In many ways, Rossetti's addition to Ferrara can also be considered the last and most splendid of the 'new lands', the seal of the thousands of new citizens who doubled the number of urban centres in Europe in the medieval period. Like many of these, new Ferrara included within it, immediately behind the walls, a clear strip of buildings to allow the soldiers to move more swiftly (Fara 1993: 47). Again, like these, new Ferrara presented an orthogonal structure, made up of right angles and straight lines. But the name which is used today for new Ferrara by its inhabitants, Arianuova, tells of a profound difference, a structural opposition similar to that which exists precisely between the air and the earth, an innovation different from all others. This is a decisive difference for the understanding of modernity, space and, perhaps too, of Tasso's fate.

Two hypotheses clash over the origins of the orthogonal schema which distinguishes the new lands. The first sustains its descendance, mediated through the medieval period, from the knowledge of Roman land surveyors, used to organizing cities (§ 69) according to the intersection of a right angle of a horizontal axis (the *decumanus*) with a vertical axis (the *cardinal*) (Comba 1993). The second, strong in terms of its discovery of the proportional basis of city plans (Guidoni 1970: 215–17), insists instead on contemporary sources: the medieval practice of trigonometry, particularly sine geometry. In theory, but also in practice, sine lines fixed the boundaries of blocks in the new Florentine medieval foundations, and their centre defined the point of intersection of the two main streets. Since sine geometry makes its official entrance into architectural literature in 1545, with the publication of the first book of Serlio's *Architecture*, the question arises: From what drawings were the architects of the first half of the fourteenth century able to draw inspiration? Where, at that time, was it possible to see geometric structures based on the circle capable of beginning the reduction of the world to space? Certainly on portolan charts, the nautical maps whose first examples date to the thirteenth century, and on the astrolabe, the instrument then in use to calculate the movement of the stars, whose flipside shows a trigonometric scale (Friedman 1988; It. trans.: 146–51, 158).

From this reconstruction, we can derive at least a few observations. The first is that the web of radial lines traced on portolan charts and the scale on the astrolabe count as visual and not technical suggestions, just like the model of new Ferrara for Tasso (§ 68), in the sense that in fourteenth-century Florence, no one was truly able to apply sines. In both cases, this is an exemplary accomplishment of one of the precepts of 'visual thinking', on whose basis the forms which surround us and which we perceive are at the root of our concepts (Arnheim 1969; It. trans.: 35). The second concerns, instead, the problem of the birth of modern space and confirms its essential nonexistence for the whole of the medieval period (§ 6). The structure of the new lands, much more regular than other medieval cities, is not yet completely spatial, despite its geometrical and symmetrical character. This is not only because of the approximation of measurements, because of the fact, that is, that nothing in these measurements actually correspond to the measurements in the plan, but is thanks precisely to their proportional character. A proportion does not equal a standard, implying only the correspondence between several elements in a reciprocal relation, in this case, the circumference's arc and its chord (Friedman 1988; It. trans.: 137–45). Quite different from the standard, the criterion of proportions in the new lands is internal to the inhabitation's structure itself, deriving from the relation between its components. In San Giovanni Valdarno, blocks became smaller in depth depending on their distance from the main street's axis towards the walls of the city. In the new lands, the unifying dimension is identified with distance, along the axis of the main street, between the centre of the main square and the end of the first block. All the other proportional values of the structure derive from this (ibid.; It. trans.: 135–7). But in Arianuova, all of this is not valid, because, between itself and the new lands, there is perspective.

70. *Arianuova*

The difference between old Ferrara and Arianuova was explained by Bruno Zevi. In the first (this is also true of all new lands), the development of the city coincides with the development of buildings, so that it would be absurd to imagine its ways as simple road circuits, 'without the third dimension of houses'. In the second, on the contrary, it is only the plane which truly counts and the two-dimensionality of the project which governs it. The architecture and the buildings acquire significance only as a function of its definition. There is no longer any common rule which is valid, as in the new lands, both for buildings and streets, dictating relationships and dimensions. Instead, building volumes must align the regularity and straightness of the arteries to slow the vanishing point, according to a sublime blend of full and empty spaces, so that wide spaces open out between buildings, such as porticoes between the buildings and the streets (Zevi 1960: 143, 158–9). The new lands possess a metrical reason which is internal and coincides with an element or relationship of the plan. Arianuova has a metrical reason which does not coincide with any element or relationship of the plan but with the reason of the plan itself, with the logic governing the set of all its relations. In this sense we should heed the claim that modern Ferrara is not a city referable to a type, a law, a doctrine, but is concretely itself (ibid.: 512). It functions as the measure of itself. It is the very first city to function as such, in that modern urban space (the reason of the plan, precisely) makes its first test there, and does so through the gaps in construction along the streets themselves, though it may seem strange. Such continuous solutions for creating depth do not simply express the progressive character of construction; their nature is not solely temporal but literally spatial, holding a structural function. These inaugurate a principle that reconfigures the indexical nature of the relation between house and street from top to bottom, abolishing its immediacy and systematic character. In this way, the conception of the city loses all solidarity and can be broken down, contrary to what happened in the new lands, into distinct phases and operations, the first being the affirmation of the autonomy of the street system against the rest.

We are speaking of the breaking of the organic link between street and building arteries, the object of Jacobs' critique (§ 60) and the origin of the contradiction between formal consideration and function of the city (§ 64). Precisely because it is directed towards the reduction of the world to journey time, the effect of this splitting was in fact, paradoxically, not the abolition of the link in question but, conversely, its generalization into the perspectival form which will mark all of modernity. This is the form founded on the advance of the straight axis and on the more or less successive and consequent expansion of buildings to infinity.

For Luciano Bellosi (1980: 23–4), the discovery of perspectival laws was a predominantly Florentine event, 'city-like', produced through the ability to see the ways among streets which function as theatre scenes where roofs, windows and cornices are arranged on different levels. This diversity nevertheless underlines a constant: that all these lines seem to lead towards a single point. The question arises whether without the Ptolemaic model of projection, which had just been rediscovered (§ 4), all of this would have been noticed. But in any case, while Florence discovers that it is a perspectival city in the course of the fifteenth century, and that it had always unconsciously been one, Arianuova is consciously constructed as such from scratch, and is arranged in a fashion much more modern and perspectival even than Florence. For this reason, its perspectives are open, different to what happens in the Florentine structure; the street scenes are never closed, the various road axes never interrupted by a frontal view, leading out to the open country, ready to colonize everything else.

Tasso brings the first modern European poem to a close between 1572 and 1577, wandering around this ghost city without houses and inhabitants, a city made only of air—the first modern city of Europe precisely because it is made of air. In doing so, Tasso becomes the first madman of modernity.

71. *In Praise of Folly*

At the beginning of the nineteenth century, Tasso becomes a typical figure of melancholy for having dared, like Ovid and Jean-Jacques Rousseau, to love a woman of too elevated a rank, and so to have become lovesick (Starobinski 1992; It. trans.: 22). Goethe (1790) gives his own contribution to this legend and cliché of the Romantic period, underlining the necessarily dramatic character of the relation between poet and society, between ideal and reality. A reality which consisted in the decline of the Renaissance, in an Italy torn apart and in total dissolution, in a fading world in which the splendour of the courts served only to mask the incipient crisis. In this Italy, Tasso was not a citizen of any city (Caretti 1961: 92), with most of the post-boys who led the carriages of the wealthy branch of the family beyond the Alps eradicated; he wore a tuft of badger fur in full view (Romagnoli 1966: 11) to feign being in a better position. The coachmen use the world from the point of view of its modernization; they are themselves the agents of its transformation in this direction. Tasso, instead, long before Goethe's *Faust*, long before Rousseau, tragically lives the terrible first experience of modernity, which makes its debut in Ferrara itself.

'Our soul is a city,' he writes (Tasso 1875: 471). His soul, just like the city of Ferrara, is not only split in two but also shows all the signs of the modern conflict, that is, the contradiction between the vertiginous metamorphosis of what exists and the destruction of everything dear to us. In the final act of *Faust*, Marshall Berman (1982; It. trans.: 98) perceives a celebration of what he calls the 'Faustian model' of development, which privileges pharaonic plans related to transport at the expense of immediate profits, in view of a consequent development of the productive forces. Between the nineteenth and twentieth centuries, this model will truly change the whole Earth: think of the cutting of the isthmuses of Suez and Panama (§ 54). The model, however, is born at the end of the fifteenth century in Ferrara, out of the destruction (Bocchi 1982) of the orchards, villas and gardens required for the rapid doubling of the city's perimeter. Modern Ferrara is, in fact, the first place of which the quotation from the *Communist Manifesto* (very dear to Berman) can be claimed to be true, a quotation Marx uses to describe capitalist development: 'All which is solid melts into air, all that is sacred is profaned, and man is at last compelled to face with sober senses his real conditions of life, and his relations with his kind' (quoted in Berman 1982; It. trans.: 120). This is the new air breathed for the first time in Arianuova, along its straight and widened roads made for inhabitants who will not arrive for centuries. Even today, Ferrara is the only old Italian city in which trees and flowerbeds reach into the centre of the city, near which, well inside the city walls, it is still possible to perceive not only gardens but also fields, stretches of authentic countryside. Tasso was very clear in his recognition of the destruction and profanation of places (§ 3) to make way for space (Farinelli 1985). His folly was not resigning himself to the necessity and inevitability of the process, not wanting to submit to another compliance whose modernizing experience was constraining: sober disenchantment about one's own conditions and relations with others.

This can be said otherwise, not yet returning to Marx but, instead, to the finest travel book ever written. Travel writing is divided into three main categories: those which speak of journeys; those written during journeys; and those written or conceived on journeys but which do not speak of journeys. Erasmus of Rotterdam's *In Praise of Folly* belongs to this last category, conceived on horseback in 1509. The first part of the book is based on a single idea: abstract reason does not exist; instead, the rationality of any behaviour depends on context. Translated into our terms: there is no kind of reason which, like the plan for new Ferrara, expresses rules that claim to count everywhere in all cases, destroying all local difference, just like the spatial norm. The sole difference, then, between Erasmus' and Tasso's folly in this respect is that the first manages to do precisely what the second does not want or manage to do: to accept the illusions deriving from men's belief in being characters bearing masks, reducing the whole world to a giant and ironic smile.

72. The 'Disenchantment of the World', Possible and Virtual

Using the idea of the 'disenchantment of the world', Max Weber referred to the elimination of magic as a technique of salvation for men, to the disappearance of sorcerers. Some years ago, Marcel Gauchet took up the expression to mean 'the exhaustion of the invisible realm' in order to define the process which animates the political history of religion. For

Gauchet, the essence of the religious phenomenon resides in an anthropological structure which is much deeper, capable of surviving by changing its outer clothing even to the end of religion itself. This is based on a 'principle of mobility put at the service of what is immobile', a 'principle of transformation mobilized to guarantee the intangibility of things' (Gauchet 1985: II, 10–11).

Tasso's tormented and mobile earthly existence, with his death at the convent of Sant'Onofrio sul Gianicolo, would seem the incarnation of this essence on the plane of existential practice. What is notable, however, is that both principles, much like the tendency of disenchantment, are necessary and sufficient to define modern perspective (§ 4, 5, 9, 34), the model of perception-representation-construction of the world developed between the construction of the new lands and that of Arianuova. It is the subject which is immobile, in whose service the winged eye moves. The metamorphosis of the eye, which concerns not only its form but also its function, serves to save modern man from strenuous tactile relations with things, which presupposes physical, direct and immediate encounters with each of them. The colonization of reality which results from this also tends towards the subject's disenchantment, by virtue of the rule in which the invisible, which previously participated uncontrollably in the constitution of the world, is now doubly tamed: either it remains compressed and concentrated behind the vanishing point, kept at bay under the form of infinity—and this concerns perception and representation—or the invisible assumes the nature of the possible on the plane of material construction by limiting the field to the virtual.

The virtual is not opposed to what is real but to what exists, to the actual (§ 9), of which the possible is the deferred replica. The possible is just like the real, static and already constituted, apart from its lack of existence (Lévy 1995; It. trans.: 6). Contrary to this, the virtual can be incorporated into what exists but can never be integrated, in the sense that something is always lacking, not only with respect to the existent but also with respect to itself: half of what is lacking is called difference and absence (Deleuze 1968; It. trans.: 133–4), which precisely assures the enchantment of the world. The world's disenchantment is completed in the modern era through the substitution of space for places, through the proliferation of the indexical relation between street and houses according to the form of Florentine linear perspective. As in every experience of the unconscious and the profound, affirmation and negation coexist in the practice of space (Dupont 1946; It. trans.: 96). This is true especially for its pioneers. It is a coincidence of history that Rossetti opens the construction site of Arianuova in the summer of 1492, exactly as Columbus was making his last preparations for his voyage. Both the navigator and the urban planner do the same thing: they affirm space while negating it. Columbus (§ 7) does so in refusing almost until the end to recognize the nature of the New World; Rossetti, in not recognizing any abstract primacy in the logic of the rectilinear and the orthogonal, in spatial syntax, seeking, instead, unsuccessfully to put these in the service of a living city. What unites the undertakings of Columbus and Rossetti is the reference to globality, to the totality of the field of action: the Earth for the first, the city for the second. For Zevi, Rossetti is the first modern urban planner precisely in having systematically referred each of his individual architectural interventions to the whole of the urban context of Ferrara.

The invisible—which still exists—assumes the form of the virtual for the pioneers of modernity for the precise reason that in their undertakings, as with Tasso, spatial logic and global logic coexist. It will only be for their successors that the invisible will assume the form of the possible, up to its point of disappearance.

73. Enchantment, Image, Disenchantment

No history, however, is rectilinear, not even a history of straight lines (Brusatin 1993). On the contrary. To make a history of straight roads would mean going over the entire history of humanity, even if limiting oneself to urban roads. Beginning from the Paleolithic era, in fact, and on the whole of the Earth, human settlements were conceived and constructed with extraordinary persistence and continuity according to the rectilinear and orthogonal schema which had only one aim: to make programmatically visible and available to the touch what was previously not, projecting the celestial order onto the order of the Earth, translating the metaphysical into the physical. As Jospeh Rykwert explains, the citizen knew that by crossing the *cardo* he walked in parallel to the axis on which the Sun turned; strolling along the *decumanus* he knew he was following the course of the Sun (1988; It. trans.: 229, 246). In this way, the city was an instrument for the decipherment of the meaning of the cosmos, and this certainty made the citizen feel inserted intimately into it. The city was, in short, the product of ritual, its form was symbolic—something in the place of something else (§ 12)—and its enchantment derived from this symbolic nature itself, precisely in reference to the invisible and the untouchable. This was an enchantment that, coinciding the form of the city and the universe, was transmitted to the whole universe and, at the same time, begun the process of disenchantment. The possibility for a third definition of the city derives from it, dynamic much like the relation it descends from, and different from the two definitions (men or houses) we have already examined (§ 66): the city defined as a giant symbol which serves memory and knowledge, a group of signs through which the inhabitants are identified with a common past through physical participation in ritual (ibid.; It. trans.: 226). This is evidently analogous to the definition Ritter made of the whole Earth functioning as a house for the education of humanity (§ 1).

To truly understand the reasons for urban disenchantment, for the reduction of the city to a group of visible things (without residue), we must keep one last observation by Rykwert in mind (ibid.; It. trans.: 7). This observation holds against every reduction of the city to a natural, physiological phenomenon: if the city really must be put in relation to physiology, it further resembles a dream than anything else. The benefit of this observation consists in the introduction of the mediation represented by the urban image in the understanding of the urban itself. In dreams, from the physiological point of view, we see things truly, so that there is no difference between dream image and material image. What is the material image, then, which corresponds to the city? If the city is what Rykwert holds it to be (referenced earlier), or if the city is essentially just the giant symbol holding the sky and Earth together (Duncan 1990), there can only be one image which corresponds specifically to it: the cartographic image, whose distinctive task and exclusive function are precisely the transposition of a reality charged with 'mastering an object which is in itself immeasurable, difficult to integrate into the visual orbit' (Wunenburger 1997; It. trans.: 67).

There is a relationship of interdependence, therefore, between the city and geographic representation, on the basis of which the former presupposes the latter. The enchantment of the city and the world depend on the consciousness of the non-exhaustive character of this relationship, from the consciousness that the immeasurability and difficulty of visual integration are signs of a difference in nature, or in ontological status, between the image and the world. So the question becomes: How and why does this consciousness cease? When does the urban image takes the place of the city and the city becomes the copy of its model? When, in short, does disenchantment begin? This is the same as asking when the indexical relationship between street and house ceases being virtual and becomes only possible: Is this really only in the modern period?

Our responses cannot be immediate. Before trying to formulate these, we must not lose the chance to define the city in a new way.

74. *What a City Is*

Without the city, there is no cartographic or geographic representation to speak of, and vice versa. We ask again, then: What is a city?

In reaction to the dominant model based on the evolutionary linear and cumulative sequence—which leads from the harvest of wild fruits to the city through agriculture and the village, and culminates with the State (§ 65)—its opposite has been gaining traction over the past years: the tendency to consider urban agglomeration as the motor force not only of the development of agriculture but also the appearance of rural villages and agricultural and pastoral life (Soja 2000: 19–49). In the middle of the nineteenth century, Carlo Cattaneo (1972) found the 'ideal principle' of Italian history in the city. Today, archaeologists, urban planners and geographers, inverting all previous reconstructions, begin to find the material origin of world history in the city, although without any pretence of totally substituting, in this way, the old theory. A much more complicated and detailed picture results from these than existed before, but also, at the same time, possibilities of simplification which directly concern the questions we want to respond to.

The canonical example, in this respect, is Çatalhöyük, the site discovered on the Anatolian highlands in the middle of the twentieth century, dating between 7,000 and 5,000 BCE and inhabited by 6,000 to 10,000 people: a number which made the cluster into the most crowded settlement yet known in the entire Neolithic world. Equally unheard of was the complexity of the social division of labour reconstructed from the remains: next to hunters and gatherers, there were farmers, shepherds, merchants and an extraordinary group of highly specialized manual labourers, artisans and artists (Mellaart 1967: 22–3, 99). Among these was the author or the authors of the only prehistoric town plan of the ancient world known today (ibid.: table 60; Delano-Smith 1987: 73–4). This is a fresco, found in one of the most ancient temples, which dates to around 6,150 BCE and which represents Çatalhöyük under the menacing mass of Hasan Dağ in eruption. The volcano is represented dimensionally, that is, as though it stood in front of the observer, but the settlement is instead represented from above, as if the eye assumed, we would say today, the point of view of the volcano. If the boundaries of settlements (very similar to those unearthed) were painted with a slightly more regular sign, we would not hesitate in defining such a plan as geometrical, rich with details but simultaneously, thanks to the overhead vision, absolutely abstract. Owing to this abstraction, the image is much more than the first true landscape ever painted, it is an act of urban self-consciousness, and consciousness about the specificity of the nature of a city organism (Soja 2000: 40). In other terms: by virtue of this fresco, since it is capable of reflecting abstractly upon itself, Çatalhöyük was to be considered a city, despite the fact that it was inhabited predominantly by hunters, farmers and shepherds.

To define the city in general, it is sufficient to generalize this idea: a city is any settlement capable of producing a material, public, and thus shared, image of the form and functioning of the world or a part of it. Consequently, every rivalry between cities is expressed, on the highest level, in the struggle for the affirmation and circulation of the images which they produce. No act testifies better to the seventeenth-century Italian urban crisis after the call from Versailles made by Vincenzo Coronelli (Scianna 1999: 123), the last heir of cartographic knowledge of the peninsula, charged in 1681 with creating for Louis XIV the great globes which no one in the country could or wanted to realize. This was a call which illustrated in exemplary fashion Italy's definitive loss of the leadership of the Renaissance, insofar as this related to the control of the world and knowledge of its mechanisms, and its transfer to the cities of continental Europe. As will be seen (§ 88), the production of specialized information is the motor of urban activity (Pred 1977: 71–4) even before the modern period. Only nowadays does it tend to assume an immaterial form, as do urban functions.

75. *Bedolina City*

The definition of the city just put forward fully satisfies the relative nature of the urban underlined by historians and allows us to avoid the paradoxical contrast between form and function which has long paralysed geographic analysis (§ 64). Further, it allows us to detach definitively from the topographic prejudice (§ 63), to the point of completely overturning our idea of the urban phenomenon, transforming into the city what until now we have exchanged for its absence.

Take the case of another famous prehistoric map, the so-called Rock of the Fields of Bedolina, a cave painting discovered in the heart of the Alps in Valcamonica, between Lake Iseo and the glaciers of Adamello and Cevedale, and dating to the middle of the Bronze Age (around 1,400 BCE): the most renowned example of the most ancient European geographic representations. Even today, the windings of the stream that descends from the mountain on which it is engraved, and the form of the fields, dry-stone walls, canals and paths that subdivide them correspond to the lines and points drawn thousands of years ago (Blumer 1964; Anati 1960; It. trans.: 108, 210, fig. 65, photo 20, 22). Historians of cartography do not hesitate to compare the visual characteristics of contemporary topographic maps to this schematic archetype, except for the buildings that, different from the map of Çatalhöyük, are drawn in elevation and then projected back onto the plane of the rock (Delano-Smith 1987: 62). In reality, these are a later addition, dating to the Iron Age, thus to the last millennium before the Common Era, when maps of this type lose their importance to the point of disappearing in the culture of Valcamonica (Sansoni 1982: 68). The canals and the planning of the exploitation of the water and land (on the map, fields destined for shepherding and cultivation are distinguished from one another, as well as four different types of cultivation) make one think of a sort of agrarian collectivism based on the clan and the tribe (Anati 1960; It. trans.: 114, 202–03). As hard as we try to complicate the social organization of the prehistoric inhabitants of the place, their settlement remains a village made up of huts. This is according to the optic of the anthropologist and the scholar of prehistory. It is clear, however, that according to the definition of the city we are discussing, the village is transformed, from the functional point of view and relative to the level of material culture of the period, into a truly urban organism. The cartographic cave art produced there authorizes this transformation. Just as the fresco in the temple is the map of Çatalhöyük, the 'Rock of the Fields' is the map of Bedolina. The only difference is that in the first case the value of the genitive is double, and signifies both the settlement which with its level of organization has produced it, and the object of representation. In the second case, the value of the genitive is single and is limited to the first of the meanings above but, in exchange, representation concerns not the settlement but the territory, reflecting precisely the function of domination and control which makes an aggregate of constructions into a true city.

In fact, the problem is even more subtle. It is not always easy to establish whether a group of prehistoric signs corresponds to a map. Normally this is based on a set of criteria or on the fulfilment of a series of conditions that are often doubtfully verifiable: (a) that the intent of the author is to represent the spatial relations between objects (here spatial is understood metaphorically and not specifically, implying only distance and farness, not necessarily a standard metric); (b) that all the signs are created contemporaneously (but in the map of Bedolina, the contrary occurs); (c) that these signs are appropriate and recurring, meaning frequent and identical within the representation itself; (d) that their number exceeds a certain minimum threshold; (e) that there is a maximum distance between these, or else that they are not excessively fragmentary (Delano-Smith 1987: 61–2, 74). Excluding the first two, arguable because of, if not contradicted by, the analysis of the most classic examples, these principles are restricted concretely to the last three, the same which are vital, come to think of it, in establishing whether several buildings form a single settlement. In this way, a further question arises: whether it is the idea of the map which depends on the idea of the city or vice versa. Or, whether the map and the city are constitutive of each other, whether they and their models nurture one another in turn.

76. *The Ideal City*

Beyond the real city and the dream city, which are the same, there is another type of city—the ideal city. This is the city which realizes the model of geometric equivalence, omnipotent for gods as for men, which Plato talks of in the *Gorgias* (507e–508b), and which he describes in the *Laws* (745b–e): a low city in which, around the circular high city (acropolis), the habitations and fields of every individual citizen are arranged together so as to be located exactly the same average distance from the centre as the habitations and fields of all the others. A city which is thus also circular, ideal both from the point of view of the form and the functioning, ideal indeed precisely in the absolute coincidence of these and for their immediate legibility. A city which, for this very reason, urban analysts never cease lamenting. There has never truly been a city like this. A simply circular city did not even exist in antiquity, with the partial exception of the semi-elliptical shape of Mantineia (Rykwert 1988; It. trans.: 108). Where from, then, is the Platonic form born? Not from dreams but from the map itself, passing through a very instructive route.

To understand this, we must begin from Anaximander and his original geographic task: the construction of the first map (§§ 13, 37, 40, 48). This meant a table, perhaps made of clay, whose round shape, which became the form of the world, recalled the shape of the assembly of warriors described over and over in the Homeric poems: a circular field and one, thus, possessing a centre, occupied in turn by the orators, each of whom, when his speech was finished, returned to his place, leaving the temporary position to the following orator (§ 0). For Mikhail Bakhtin (1975), the model of the centre and of the circle characterizes the epic as a literary genre, the great literature of the classical era. For Jean-Pierre Vernant, it is precisely this model which distinguishes the Greek city from other cities of the classical world, all lacking a central square (from the Phoenician cities to Babylon), which the Greeks called the *agora*, the place of public debate and thus the origin of every political institution. In the free debate which unfolds inside it, the participants are defined as 'equals', 'fellows'. A society is thus born in which the relationship between citizens assumes the form of a relation of identity, symmetry, reversibility, equilibrium, reciprocity: qualities that are the roots of the system we still call democracy (Vernant 1966; It. trans.: 210–12). We owe the most suggestive description of this city, the city of isonomy and the equality of peers before the law, to Jorge Luis Borges (1985). Two Greeks, finally free from myth and metaphor, forgetting prayers and spells, converse and agree on just one thing: that we can reach the truth through dialogue. This conversation presupposes a central location, involves the practice of debate and produces something very precious: information. Mike Davis recounts that, arriving in Los Angeles to take care of business and disoriented by the non-homogenous functioning of the immense urban aggregate, Japanese people sometimes ask: Who governs here? (1990; It. trans.: 87–124) In the case of the original Greek *polis*, we could only respond in one way: No one but, rather, one thing— the mechanism.

To summarize: the model of the city descends from that of the assembly and the centre of the first equals the centre of the second (the *agora*). From this translation into material terms of an immaterial schema is born a new urban plan in the Mediterranean, in which all houses are oriented in the same direction. Until this point, geometry was not involved in the process, if not as a reflection of it; the centre and the circle are elements deriving from the custom of assemblies, becoming an urban model—and, with Anaximander, the model of the world and the universe. In Anaximander's time, social practice produced the formal schema of knowledge and, at the same time, the material figure of the city: the archetype of the assembly is concrete and descriptive. Two centuries later, with Plato, the circular model becomes, instead, abstract and prescriptive—that is, geometrical—and is applied to the description of a non-existent city. Why? What had happened in the meantime to justify this inversion of the world and its simulacrum?

77. *City, Territory, Democracy*

In the meantime, something decisive had happened: towards the end of the sixth century BCE, the first geometric city was born—the Athens created by Cleisthenes, without which Plato would not have been able to appeal to the circular figure as a technique for the realization of a social project. In Plato's time, Cleisthenes' city, rational and homogenous like the geometric vision of the universe belonging to Anaximander (Lévêque and Vidal-Naquet 1964: 123), already ceases to work, with the result that in Plato's recourse to an ideal model, one can also detect a kind of regret. In any case, both Anaximander's representation and Cleisthenes' isonomic city assume the form of the circle, exactly like the Ionian coins of the time and the maps mocked by Herodotus (§ 40). The relationship between the first and the second is imposed according to the schema C-M-C, city-map-city, meaning: the *polis*-map of Anaximander takes as model the *polis-polis* of Cleisthenes which takes as model Anaximander's model. In different words: Cleisthenes' concrete political construction derives, before Plato's ideal construction, from a model which is not concrete but by nature abstract—that is, product of the transfer of a form from its original environment to another. Cleisthenes' *polis*, Athens between the sixth and fifth centuries BCE, is built in fact on the equivalence which Anaximander first established in the Greek world: the equivalence between world and (geometric, that is, cartographic) image of the world, to speak in the language of Heidegger (§ 5). This was two millennia before the era Heidegger himself fixes as the birth of the equivalence in question. Cleisthenes' reforms are in fact its first, powerful demonstration and, for this reason, the birth of the modern concept of political identity is the first formidable result of these reforms.

Before Cleisthenes, the identity of the citizen derived from his descent and ancestry, and from his belonging to a specific community of worship. The political game was based on factions, on a pyramid of members relating to an aristocratic peak. But with Cleisthenes, starting from 508 BCE, the basic administrative unit of Attica became the *demos*:

a village, a settlement, a district of the city, whose functionaries, democratically elected, substituted the most important family at the local level (Forrest 1963; It. trans.: 191–203). Consequently, political identity was literally re-founded on other terms, and, even, for the first time on the basis of terms at all, that is, of boundaries. Identity began to depend, above all, on the belonging to a given territory and, at the same time, on the recognition of one's position within a hitherto non-existent plan. This plan for the first time came to be interposed between the social and the political order. Society, with all its inequalities, remained what it was: women were women, slaves were slaves, foreigners were foreigners, and none of these people enjoyed in any case full rights or participated in the assembly. Next to society, however, another level was born, within which the dependence on social links was eliminated, and the nobles and simple citizens (male and propertied) were for the first time all equal, despite their inequality. This plan, which is the realm of freedom in political terms, has nothing to do with the realm of necessity constituted by the totality of social relations but is superimposed on it (Meier 1980; It. trans.: 263). Thus, in travelling between one's home and the central square, the privileged place for the exercise of rights connected to citizenship, the Athenians each time crossed an abyss (Arendt 1958; It. trans.: 25), even if the path was absolutely flat. What they did in reality was to ascend and re-descend while remaining the same person, the unbridgeable gap between social difference and political equality.

But where is the possibility of this ontological modification of one's condition born from; what is the nature of this new equality, of this generalized yet exclusive identity? Where does the new level originate from, the new order, whose advent coincides for Herodotus (*Histories*, VI.131) with the advent of the democratic regime? Where does the autonomy of political reality arise from, with respect to social reality? In short: Where is democracy born from, if not from the same cartographic grid which births, together with the first geometric city, the territory?

78. *The* Nomos *of the Map*

Isonomy implies, however, the fact that within the city the problem of power is no longer resolved by a human being but by the functioning of institutions themselves. In this way, we substitute for the person an impersonal mechanism (§ 76), a *nomos*; Carl Schmitt (1974; It. trans.: 54) has noted that, before Plato, *nomos* was a word whose original meaning was indissolubly linked to space and signified the first measurement from which derive all other criteria of measure. In truth, Schmitt maintains the strict derivation of the term from the concrete occupation and repartition of land, but this is a meaning evidently lost in the fifth century BCE. Isonomy is, at least from Cleisthenes' time onwards, a purely political notion, quite distinct from *isomoiria*, a word which at least from Solon's time onwards specifically indicates the subdivision of fields (Lévêque and Vidal-Naquet 1964: 31). What derives from it is something hard to recognize even today: the decisive function of the map in the transformation of the meaning of the term in the generic sense of law, regulation or rule made or issued, and in its political sense. Which equals the affirmation of the geometric, and thus cartographic, nature of political law itself, of the political order, that is, literally of the urban order, on which Cleisthenes bases the identity between civic environment and territorial structure. Based on this identity, power comes precisely to be confused with the surface of the earth (§ 16), and the territory is defined as an area through whose delimitation and domination, a political subject seeks to determine phenomena, human behaviours and relations (Sack 1986: 19).

At the time of Cleisthenes, knowledge becomes technique, becoming liberated, that is, from all magical and religious elements and becoming rational (Vernant 1966; It. trans.: 318). But this, which happens within the first city whose internal distances are geometrically measured (Lévêque and Vidal-Naquet 1964: 21–22, 78), would never have happened without taking Anaximander's *hubris* as a model, as well as the image of the universe and the world which resulted from it (§§ 37, 40).

This overturns the commonly held view of the origins of Cleisthenes' revolution: it occurs not simply when new realities can be inscribed on a map (ibid.: 13). On the contrary, precisely because the map becomes the machine of each construction after Anaximander, new realities and new ideas can arise, including those of Plato. Technique is not the origin of the map but the map is the origin of every technology (§ 0). Within the old *polis*, thus before Anaximander and still in his time, the equality of citizens derived from the fundamental homogeneity of citizens, who were *homoiotes*, that is, fellows, since they were related or resembled one another from the qualitative point of view. Contrary to this, the isonomy affirmed in the fifth century BCE is based on quantitative equality, because it designates 'something which can be divided into absolutely equal parts, like spoils' (Meier 1980; It. trans.: 301, 303).

We will soon see what precious spoils we are speaking of. Note, in the meanwhile, that within the Homeric assembly, as within the *polis* in the first centuries CE, there is no space, because the relations between subjects and objects flee from all abstract and standard measurement, and all distance fixed by impersonal rules. What is the distance between the orator and the other warriors in the assembly? It is a radius whose length obeys, like the circular form of the gathering, after all, the need to ensure the most equal distribution of information to all participants. Just as when a group of people is seated around a fire to warm themselves, the circle is the figure capable of assuring to each of the members the most equal quantity possible of what emanates from the source. And in the assembly, distance depends, like the distance between Ulysses and Polyphemus on the act of the first shout (§ 59), on the relation between two physical functions: the voice of the orator and the ears of his peers. Thus, the Athens of Cleisthenes is the last Greek city made on a human scale, because in it, which begins to take the map as its model, the human measure begins to be overwhelmed by the abstract spatial metric.

79. Herodotus' Laughter

Isonomy is the product of Anaximander's operation and of the derivation of the map from urban form. Since it is on this operation that Cleisthenes bases the civic environment (§ 77), it is still differentiated from the qualitative point of view (Lévèque and Vidal-Naquet 1964: 77). Pericles' democracy, half a century later, emerges from the overthrow of this relation, thus from the prevailing of the cartographic image in relation to urban reality, from the modelling of the latter on the former. To understand the reasons for this, we must summon, after Herodotus, another of Pericles' friends, Hippodamus of Miletus, the first great urban planner and architect of the Greek world, and also, correctly speaking, the first political theorist. Today, his name is remembered, above all, as being the author—although it would be better to say the coder—of the orthogonal urban plan, in which streets cross one another at right angles within a quadrangular space: the grid layout (Morachiello 2003: 9–69).

It is no coincidence that Herodotus, born in Halicarnassus, proudly proclaims himself in the first line of his work as the citizen of a fellow city: citizen of Turi, the pan-Hellenic colony founded in 444 BCE in southern Italy to assure the merging of eastern and western Greeks. Turi was a city which incorporated the image of the world functional to Athenian imperialism and its project of the creation of a common Mediterranean market within which even the antithesis between Greek and barbarian came to be negated. Another of Pericles' friends, Antiphonus, testifies to this when he affirms that by nature foreigners and Greeks were all equals (Diels and Kranz 1922: 87). Pericles' power politics was not only based on simple territorial expansion, it was also founded on the spread of the Athenian vision of the world (Nenci 1979: 45–6), relying, that is, on the exportation and imposition of exemplary models: starting precisely from the most striking and impressive model—the urban model. Free from all past urban inheritance, all historical conditioning, different from Athens, Turi was what the Athens of Pericles should have been, materializing its ideal type or, anyway, coming as close to it as possible. It was thus not circular in form but obeyed an unheard of schema, at least in Greece—it was rectangular.

The genuine reason for this form will be revealed three centuries later, more or less consciously, by Geminus the Stoic, who in his *Introduction to Phenomena* writes: 'We must be wary of the distances signalled on round charts, since the *ecumene* cannot be circumscribed by a circle, being a portion of the sphere whose length is double its width' (16.5). Since Western knowledge of the world followed the border of the Euro-African Mediterranean in its development, it had been long realized that this was more developed from west to east than from north to south (Aujac 1987a: 135). In fact, all the maps whose description Herodotus provides in his *Histories* are already quadrangular, constructed on the basis of a linear axis which coincides with a means of communication, as in Persian custom (Myres 1896: 628). This is the reason why he laughs at Ionian maps which he finds to be so perfectly round as though coming always from the hands of the same potter (§ 40). This is because, as Geminus precisely reveals, the problem of circular maps consists in the fact that the more one moves away from the centre, the more the relative distance between two points (the first piece of information required for the organization of a functional market) is inevitably distorted. On the contrary, this does not happen if the world assumes quadrangular form, as on the maps Herodotus uses, because in this case, even the points which are found on the edges are, if revealed, separated by a rectilinear segment. And is it not still today the map that is the model of the world which sacrifices, because of its faith in the linear distance between two points, every other piece of information?

Precisely as Pierre Lévêque and Pierre Vidal-Naquet (1964: 133) maintain, the new vision of the political order, from Hippodamus to Plato, is no longer based on the city itself but 'on the order of the world'. Correct: but only if we take it that 'world' here stands for map, that the identity between world and map has mutated into the domination of the map—of space—over the world.

80. *The Squaring of the Circle*

To understand what order this is, we must look to Figure 1 which represents the ideal schema of the isonomic city, and put it in relation to the triangle created by Frege to illustrate the difference between sense and reference (§ 13). In the circular city, senses are innumerable, and the vectors of sense, the streets—that is, the radii whose paths correspond to the hypothetical trajectory of citizens directed towards the centre, into the square—are by definition absolutely identical. The distance between two points (between the habitations of two citizens) increases as one moves from the *agora* towards the periphery: the segment *ab* is smaller than the segment *a2b2*. For this very reason, similarly, Plato espouses the model described earlier (§ 76). But what remains decisive is the fact that in relation to the centre itself, meaning the public space, the distance is absolutely identical in both cases, and this is the best definition of what is concretely to be meant by isonomy. This is also essentially the best illustration of how the square is the place in which the multiplicity of senses is transformed into reference (§ 76).

Look now to figure 2, the quadrangular schema of the Hippodamian city: the relation between centre and citizens, and between citizen and citizen, is overturned with respect to the preceding schema. First we must clarify that which does not result from the schema: while the geometric centre, seat of all the most important activities, coincides with the functional city in the circular city, this coincidence is lacking in the quadrangular city, which is more complex, and the temple, the *agora*, the market are dissociated and dislocated to different points. In all cases, the distance between individual citizens remains identical, independent from the distance from the centre: the segment *ab* is the same as the segment *a2b2*. A genuine opposition is in this way delineated: within the circular city, the distance of each citizen from the centre is equal, but the distance between citizens is unequal; within the quadrangular city it is the opposite, equal between citizens, unequal from the centre. The passage from isonomy to democracy is precisely realized in this inversion.

The only advantage of democracy concerns the speed of the circulation of information, which in the Hippodamian schema is much quicker than in Cleisthenes' city, owing to citizens' equidistance from the centre. Information is the precious spoil which we must split (§ 78), and its circulation becomes faster only because the distance is standardized between citizens, and only because the city plan becomes, for the first time, fully spatial. For Pericles, democracy is isonomy plus speed of information, assured by the rectilinear and orthogonal axes, according to the principle regulating the transformation of the entire Mediterranean into a single market: the principle of straightening, of the reduction of the world and its cities to a giant quadrangular map (§ 79). Cleisthenes' city, ideally circular, is still the city in which the citizens are equal in front of the law and participate in decisions, in the centre which guarantees, with its uniqueness and geometric character, the coherence of the urban order and the order of the cosmos (§ 73). In Pericles' city, in which this coherence is lost and the centre no longer exists in its uniqueness, information is faster but decisions are no longer collective, and few know what is actually happening. As Thucydides explains: 'The government was a democracy in name but in reality ruled by the first citizen' (*History of the Peloponnesian War*, II.65.9).

Pericles' project failed, because it was impossible to find a solution to the problem which is ours even today: to reconcile the grounds of democracy with the functioning of the market. This is later, even if no one any longer remembers, the true question hidden behind the problem of the squaring of the circle, the construction of a quadrate whose area is equivalent to that of a given circle, that is, to the straightening of its circumference. A problem which is only geometrical in appearance but in reality is the story of two cities irreducible to each other: the circular city and the quadrangular city, the city of isonomy and that of space (Farinelli 1994).

81. *Reflection on the Baroque*

According to Jean-Pierre Vernant, Hippodamus still tries, in one attempt at reflection, to hold together physical environment, political environment and urban environment (1966; It. trans.: 259). This is an open question. Certainly, however, the city ceases to be a work of nature and the gods and becomes, through its plan, something entirely different. Long before Hippodamus, for example, Babylon already had a grid layout and in it the streets crossed one another at right angles. But it is very probable that it was Hippodamus who would change the value and function of this, stripping it, together with the straight line, of all religious symbolism and entrusting it with a contrary mission: to impose the order of reason within the civic environment. Aristotle explains in *Politics* that Hippodamus invented the geometric layout of cities. As Jean-Bernard Racine wrote, he 'established human residence in a new land, the land of mathematics' (1993: 141). In this way, he desacralizes the city and, at the same time, organizes it into a system so that the urban schema becomes the result of a calculation abstracted from all externals, to the form of the city itself and its functioning. Already in 434 CE, the colony of Turi no longer recognized Athens as metropolis, as mother-city, signalling the imminent end of Pericles' dream: to export a formula for government based on the most radical and direct democracy through power politics. But again in the modern period, from the Spanish settlements in South America to the Dutch and English settlements in North America, the Hippodamian plan will mark the advance of Western colonialism in the New World and the advance of the Enlightenment city in Europe: think only, in this regard, of the borough constructed in Bari at the beginning of the nineteenth century, which is still named after Gioacchino Murat, the king of Naples at the time of the Napoleonic conquest of the Italian peninsula.

It was precisely the grid layout of the Murattian borough that would prefigure the urban renewal programme beginning in many Italian cities after unification (De Seta 1976: 406). All urban planning of the eighteenth and nineteenth centuries, however, just as in the following century, depends on the nature of the Baroque city, which represents the peak of the identification of the city with space, and of the organization of the urban according to the cartographic principle. It is the Baroque city which will take up and develop the operation began by Hippodamus even to its most extreme consequences: the transformation of the city into the map of itself. Lewis Mumford's analysis of the regulatory Baroque plan is still classic (1938; It. trans.: 115–22). He argues that the city is sacrificed to the artery, that is, to traffic; that the abstract geometrical schema determines the social content, in the sense that it precedes the needs of life and conditions the institutions of the community; and that if the terrain chosen for the foundations is irregular, it is flattened (evidently because, just as a set of buildings is modelled on a pre-existing design, in the same way, and already before this, its base is modelled on the base of the design itself, that is, on the table). What derived from this, moreover, thanks to the excessive symmetry and rigidity of the plan, was the expulsion of any temporal dimension, of any adaptation of urban formations across time to the needs of successive generations. The Baroque city had to be constructed just as it had been projected: in one piece, and under the guidance of a tyrannical architect. As Gilles Deleuze reiterates, the city becomes a table of information (1988; It. trans.: 41).

It is in this city that the nature of the indexical relation between street and house becomes irreversibly possible, and no longer virtual (§§ 60, 72, 73), assuming, that is, the form of the spatial relation. The city becomes a diagram within which the existent has already colonized all the forms of the future. This occurs because in the Baroque city, the straight roads perform exactly the same function as maps, as Robert Harbison realizes (2000: 60–4) at the crossroads of the Quattro Canti in Palermo: they excuse us from direct experience of the contorted surrounding reality, allowing us to see everything immediately, to access instantly, like in a good filing system, the part which interests us. Hence, between the seventeenth and eighteenth centuries, it is thus through the mediation of streets that cities become, like the Earth itself (§§ 5, 9), the copy of their own copy.

82. *Imaginary Space*

Consider that the Baroque arteries are, from the functional point of view, contradictory with respect to their form: these lead in a straight line to the heart of the city but are actually equal circumvolutions whose goal is to avoid the friction of urban crossings (§ 9). Behind this contradiction, the powerful invention of imaginary space is active, indispensable for the understanding of modernity, and which we owe to Thomas Hobbes, the philosopher to whom we also owe the reflection which accompanies and engenders the birth of the territorial, centralized state.

For Hobbes, the world is indeed a map—to conceive of something means to conceive of it in some place, and as possessing a certain extent, divisible into parts and such that it cannot exist at the same time in two different places, just as two things cannot be in the same place at the same time. The only science that God has given to humanity is geometry, and it is with geometry that man begins fixing the signifiers of things, which men call definitions. This fixation corresponds to law, in the sense that the intention of law—its meaning—is its literal sense, which is one and only one (Hobbes 1951[1651]; It. trans.: 99, 105, 326). This is exactly as on and as only on the map, where all names are proper names (§ 17, Farinelli 1992: 3–14) and where exists a bijective correspondence between name and thing: 'A name means an object. The object is its meaning,' as Wittgenstein writes in *Tractatus* (1922: 3.203). It is on the basis of such criteria—on the basis, that is, of the cartographic image—that the Baroque period begins the material construction of the state as artificial product of human calculation and the mechanization of state representation. This is a decisive step, from which the rest of the modern world, 'from drainage to the chemical process, consequently descends, and has no need of any further metaphysical decision' (Schmitt 1938: 53, 59).

For Hobbes, scientific knowledge was not originary; it did not concern the ontological field and it had nothing directly to do with the structure and the nature of the external world. On the contrary, it derived, in its possibility, from the cancellation of the world. The subject of scientific knowledge was conceived as a man for whom, with the world having disappeared, there remained only ideas and images of things previously seen and perceived. Thus, he could do nothing other than calculate on the basis of his own ideas, his own phantoms, as though these were external and not generated by his mind. From here came the imposition of schemas onto the world, as a function of its control, which were artificial because they were based on an interior criterion of validity. The reappearance of the world at this point should not surprise us: Hobbes' hypothesis was not at all existential. It did not concern the real existence of things but was only the illustration of the modalities of the construction of knowledge. The schema obtained through the annihilation of external reality was imaginary space, an unreal extension, a phantom image of what really exists. It was the abstraction of that dimension in which bodies, removed from their accidents and properties, are presented simply as external objects, an artificial system of points and positions conceived as the sole means for understanding the behaviour of physical objects. In this way, imaginary space allowed one to define the movement of bodies which are not at all imaginary, because it allowed one to define the place, that is, the portion of space which a body coincides or is co-extensive with (Gargani 1971: 138). As Hobbes writes: 'place has no existence anything outside of the mind; physical size has no existence inside the mind' (1951[1651]; It. trans.: 379). Inside the mind there is a 'carte blanche', a *tabula rasa*. Outside this, on the plane of originary knowledge or fact, bodies are presented according to a law and an order which obey a system of interactions that are mechanically determined, as Francis Bacon had established (Gargani 1971: 131): thus, objects already assume a cartographic form; so that for Hobbes, knowledge is already arranged according to the model of mapping (§ 37) and the relation between two maps. Precisely because the interior one, which corresponds to imaginary space, prevails over the external one and subsumes it, the artificial rectilinear form assumes the entire functioning of the world upon itself.

83. *The Result of Modernity*

The result of modernity consists thus, in fact, in the reduction, through mapping, of the world to a map, to a table. Through this process, the unreal mutates into the real, the face of the Earth is transformed into the imaginary space of Hobbes, assuming, that is, the features of Euclidean extension, in a surface which obeys the rules of continuity, homogeneity and isotropy (§ 4). The modern territorial state is the result and the organizing agent of this transformation, whose concrete articulation develops according to the succession: straight road–railway– motorway. The first, whose creation is the mature product of the Baroque period (§ 5), functions as model for the railway which in turn does so for the motorway. In Germany, the motorway is still called the *Autobahn* which means 'railway for automobiles'. If the advent of straight-road routes corresponds to the original stage of state formation (§ 9), the development of the railways coincides with its maturity, and that of the motorways with the beginning of its decline.

Straight roads obey an artificial model in their form, but the animal-powered vehicles which, up until the first half of the nineteenth century, cross over these roads still respond to the logic of natural movement. Only the mechanization of traffic on land (and on water), owing to the employment of steam power, overturns the relation between natural fact and means of locomotion. With it, the speed and characteristics of movement no longer depend on the former but on the latter, because mechanical power constructs its own environment emancipated from every feature of nature, and which coincides in all respects with the result of Hobbes' operation: a body precisely deprived of all qualities and reduced to a pure, abstract extension as a field of mechanical relations.

If Hobbes spoke of 'annichilatio mundi' (Gargani 1971: 138), the first records of the spread of the railroad speak of 'annihilation of time and space' (Schivelbusch 1977; It. trans.: 11). In reality, this confused what was being born with what was dying, because it is to the railways themselves that we owe the definitive transformation of the world into space.

Wolfgang Schivelbusch explained it very well (1977; It. trans.: 21–2): the steam locomotive produces mechanical, uniform movement and the unit of wheels and rails transfers this movement to the Earth's surface. Thus, the railway is the technical means for the application of Newton's Laws of Motion on the surface of the Earth, in which every body perseveres in its state of stillness or uniform, straight movement if no force applied to it constrains it to change. What follows is that if it were possible to construct an absolutely smooth, flat, hard and straight road between two points, to move a vehicle on it would require only the force of traction necessary to overcome the resistance of the air. In other terms, it is precisely with the railway, which is the model of the ideal road because it is without friction, that the mechanization of movement transmits the decisive attribute for its translation in spatial terms onto the skin of the whole Earth: the standard. This is the case from the extensive as from the intensive point of view. Vidal de la Blache figuratively describes the essential and indissoluble relation between the railway and the process of colonization of the American continent and between the railway and the formation of European national states (1922: 247–50, 253–6). In fact, the system of railroads and the modern State function according to exactly the same principles, behaving like a great machine and requiring unitary direction and coordinated movements, precisely because both are agents and, at the same time, products of the spatial model: both presuppose a continuous, homogenous stretch in which all points lead towards a centre.

Counterpoint, however: the crisis of the spatial model involves both the crisis of the state and that of the railway. Let's look at a single comparison. In 1840, the length of the railways did not reach 8,000 kilometres over the whole of the Earth; it became 206,000 kilometres in 1870, 790,000 kilometres in 1900 and more than 1,300,000 kilometres in 1911. This last figure corresponds to more than 25 times the circumference of the globe (ibid.: 244) and equals the size calculated today on the basis of the data supplied by the International Union of Railways.

84. *Once There Was the Sea, Once There Was the Earth*

In reality, the last piece of data does not allow us to distinguish between the total length of the railways, that is, of the tracks, and the length of the system, which in many stretches contains more than one track. In any case, the momentum of the railways ends on the eve of the First World War, precisely because of its development. Vidal explains (1922: 250) that it was much before this that we began to realize that the railway revolution consisted more in the movement of things than of travellers, and this is undoubtedly the case, although at the outset no one thought about people (Capot-Rey 1946: 98–9). At the same time, it was precisely through the progress of the railway routes that things began to be transformed into immaterial impulses and to be moved from one point to another much faster than the locomotive itself could make them. It was the railroads that led to the birth of telecommunications and, with this, supplied the first leap towards the dematerialization of the world (§ 24) and, consequently, to the process which only recently, and in another context, found its most precise definition: 'despatialization' (Abu-Lughod 1999: 272). It was the railways that encouraged and guided the advance of the electric telegraph.

In the United States, their spread occurred, starting from the 1830s, through a kind of symbiosis. The telegraph allowed the railroad track to avoid incidents, as well as track doubling, because it allowed convoys travelling in the opposite direction to avoid collisions without having to observe long stops at stations, and waiting, instead, only for the necessary amount of time (Hugill 1993: 320). It thus saved time and money, which became definitively the same thing in this very way. It was thus that the model of the railway system was transferred to telecommunications and, already in the first half of the nineteenth century, was codified in relation to the concept of the network understood as 'the interweaving of objects arranged in lines' (Mattelart 1994; It. trans.: 68): a definition which explains better than any other the reduction of the Earth's surface to its cartographic image. In 1855, the laying of underwater cables began and Great Britain, which constructed the first global system of information in this way, maintained the hegemony of communication for around a century, capable of linking all lands, seas and oceans apart from the polar ones. It was only with the laying of the first Transatlantic telephone cable, finished in 1956, that this hegemony began to transfer to the United States (Hugill 1999: 27–51, 228–31).

This is how land and sea, stripped of all accidents and qualities (as per Hobbes), became exactly the same thing, at least as far as they concerned the transmission of the most precious goods: money and information, the most important goods for the functioning of the world. This thing, however, is no longer just space, precisely because this functioning is less and less attributable to a metric, or to a problem of speed. Stephen Kern (1983; It. trans.: 157) reminds us that the psychiatrist who introduced the diagnostic category of 'neurasthenia' into the literature maintained that, with respect to the eighteenth century, the telegraph, the railways and steam power had allowed the number of transactions made in a given period to multiply a hundredfold. But the railways and the telegraph, which indeed collaborate in the construction of the new world (and its specific illnesses), do not function, for their part, in the same way. The telegraph too, like the railways, sees an increase in speed between the nineteenth and twentieth centuries: from a transmission capacity of 25 to 28 words per minute to more than 100 at the beginning of the last century, and to more than 300 in the 1920s (Hugill 1999: 35). Different from the railways, however, the telegraph changes the nature of things, dematerializing a printed message at its starting point and making it reappear as such on arrival, separating the circulation of information specialized in human interaction from the concrete relation between people. That is, it transforms what exists, which we can touch and count, into what *subsists*, which we can think but neither touch nor count. This very split and division of labour (what exists being the railways and, more generally, the means of material communication, and what subsists being electronic transmission and immaterial communication) is the origin of the crisis of space.

85. Les Demoiselles de Midi

The fame of the Midi girls dates at least to Guido Cavalcanti, a friend of Dante, who describes one in a sonnet his devout pilgrimages from Florence to the sanctuary of Santiago de Compostela, prematurely interrupted at Toulouse thanks to the grace of its inhabitants. The girls of Avignon gave the title to Pablo Picasso's 1907 composition which is usually traced back to the crisis of the modern pictorial image in terms of the decline of traditional spatio-temporal referents: the painting disappears as an imitation of objective reality, of nature, since the horizon line disappears, as does depth (the distinction between different levels, the perspectival model) and, therefore, the unity of the subject and the coherence of the object (Lefebvre 1974; It. trans.: 292. See also Kern 1983; It. trans.: 367–96). This occurs because cubism first transforms the painting into a map.

The Customs officials who blocked Igor Stravinsky at the frontier between Italy and Switzerland in 1917 realized this, without knowing why they were right, and accused him of wanting to smuggle a map, because they found his portrait drawn by Picasso in his baggage (Stravinsky 1935; It. trans.: 66). In it, as in cubist works more generally as well as on maps, the picture is made up of a single surface which completely fills it; the third dimension is also reduced onto this single plane, and the height of objects is reduced to a series of geometric forms. *Les*

Demoiselles d'Avignon caused an enormous sensation on their appearance because they had distorted noses and eyes, as though observed by a subject who simultaneously looked from two different points of view. It was precisely the consequent questioning of the unified point of view and the immobility of the subject (§§ 4, 34) which caused this scandal. All this would not have happened, however, if Picasso had not first silently applied to the frontal image on canvas the same rules which the cartographer, whose eye looms static at 90 degrees, applies to the map. This is another half-turn (§ 58), this time only of the eye, which for modernity is normal (§§ 5, 9, 11) and, for this very reason, scandalous. With this half-turn, we leave modernity (the world reduced to space) just as we entered into it with the turn of Ulysses which concerned the entire body. The flattening both of the object and of the subject depends on this turn, which relates to Picasso as the first painter whose production is understandable only by making reference to a series of distinct 'periods' or multiple and different expressive styles. This flattening depends in turn on the collapse of the central and decisive element of the system: subject–standard linear metrical distance–object, which crowned the flight from Polyphemus (§ 59). This system was, between the nineteenth and twentieth centuries, undermined more by the telegraph than the locomotive in terms of the overall mechanics of the world. The cubist movement, too, undermined this system in terms of the use of the artistic work, because it constrained us to look at a map at the same distance from which we look at a picture.

At the same time, Picasso disarticulates the surface itself of the picture-map, as well as objects, precisely because these are the years of the crisis of the spatial schema, which result from the imposition of the cartographic model on the world. It is in this sense that Picasso supplies the vision that the world was waiting for, as Henri Lefebvre maintains (1974; It. trans.: 293). The significant is separated from the expressive, the sign (the signifier) is separated from what it designates (the signified), and the sign is no longer the 'object' but the object on canvas, thus becoming the treatment that the objective fact, decomposed and reduced, submits to. This is exactly what occurs to words in the course of telegraphic transmission, with the only variant that the telegraph uses paper in place of canvas. In this way, then, we are indeed speaking of an annihilation of space, as was erroneously said in relation to the railway network (§ 83), and of a destruction of space's characteristics of continuity, homogeneity and isotropy. So that the same fate exists for space as was reserved for the concept of 'life-force' in the most mysterious and allusive of Humboldt's writings (1849): the day of its triumph (the railways in this case) is also the day of the start of its death (the telegraph).

86. *Metropolis: From the Railway to the Motorway*

For the Ancient Greeks, metropolis meant mother-city and implied a relation with a child-city, with a colony, such as Athens in relation to Turi (§ 79). Here we mean by metropolis the city where the triumph and simultaneous death of space is recorded, meant not only in the general technical sense employed from the outset (§ 3) but also in the specific form of the indexical relation possible between street and house (§§ 72, 81).

On the American continent, as in the Brazilian *sertão* which is opposite to the coast of the ancient settlement, the railway was given the name 'planter of cities' (Deffontaines 1938: 323), generator of urban realities from nothing. In fact, it has played a decisive role in the growth and extension of all the great agglomerations which have appeared in the last two centuries in the temperate zone of the two hemispheres. It is to the railroad that we owe, within those, the development of the peripheries, in a process which everywhere exhibits the same characteristics and is essentially based on the necessities of provisioning, before the appearance of the automobile at the beginning of the twentieth century (Capot-Rey 1946: 254–8). Perhaps the most obvious example in this respect is constituted by the formation of Greater Berlin, founded in 1920 by the fusion into a single administrative unit of the city of Berlin with 7 other cities and 85 surrounding rural municipalities. Not only did this fusion take on the directions of the existing rail routes but also the real-estate firms interested in the subdivision of peripheral areas linked these to the centre with new lines which were, significantly, called 'colonization lines' (Le Roy 1935). These conserved the memory of their original function in the name, which is still reflected in the model of current metropolitan systems used throughout the world. Berlin's was not the first metropolitan railway: London began to construct its own in the 1860s; the first section of New York's was opened in 1904—and both managed, as in the case of Berlin, to anticipate and lead urban development, welding together pre-existing nuclei of settlements. In general, already in the first half of the twentieth century, the railroad functions as an index of urban settlement, in the sense that it draws and directs it. In the development of today's metropolitan systems (just think of Los Angeles), on the contrary, urban settlement precedes the railroad which no longer succeeds in channelling urban growth.

This inversion, which is much more significant than it seems, passes through the complete breaking of the link between road axes and urban structure, signalling the complete loss, therefore, of the functional meaning of the spatial model, owing to the arrival of motorways. Motorways are streets that programmatically do not allow one to enter any building. In Europe, a continent which has been rich in cities for centuries and millennia, the motorway serves to avoid the city, allowing one not to waste time (§ 9), different from what occurs in the case of the railways which strengthens the link between urban settlements and streets (Gambi 1984: 136). The first motorways were constructed in Italy, Germany and Holland in the interwar period, and spread in the postwar period to the rest of the continent. The nature and the affair of the American motorways is, instead, very different, and the difference from the European motorways is an indication of the profound difference in their relations to urban centres. In the United States, homeland of individual motorization, the motorways—and not solely the railways—bore the task, through the urbanization of the peripheries, of transforming the city into a metropolis in the twentieth century. In 1911, the Long Island Motor Parkway was opened in New York, the first motorway in the world with limited access, constructed to allow pedestrians to reach offices in Manhattan from their residences situated between a radius of 50 to 55 kilometres. In the 1950s, the first freeways appeared in Los Angeles, urban motorways which, substituting themselves for public lines in the transit of goods on the tracks, transformed the routes into their exact opposite—that is, into so many large walls, barriers erected between rich and poor areas (Abu-Lughod 1999: 198, 253–4). With streets for cars which separate instead of unite, space truly begins to end, just like the city, and the metropolis begins to construct its colonies within itself.

87. *Mesopolis: From the Street to the Railway*

Instead, in Europe, land of ancient settlements, there existed urban groups supportive of one another long before the telegraph and the automobile. These were constituted by a series of more or less large cities interdependent in such a way that every significant change in economic activities, in employment structure, in revenue or in the demographic consistency of a unit was reflected directly or indirectly in the others. One of the most extraordinary historical examples is represented, in Italy, by the 'unrepeatable bimillenial phenomenon of the via Emilia' (Kormoss 1978: 42). This was the alignment of city centres along the Apennines, almost in a straight line from Rimini to Piacenza. Together, these constitute a 'conurbation', in the sense that Patrick Geddes, inventor of the term in 1915, assigned to it: not an area urbanized seamlessly (Fawcett 1932: 100) but a 'galaxy of cities', a 'natural alliance of cities'— in short, a 'city-region' (Geddes 1915; It. trans.: 57–61).

In the case of Emilia Romagna, the last expression should be understood literally: the via Emilia is a gigantic *decumanus* (§§ 69, 73) which divides the region into two halves—the plains and the mountains (Nissen 1902: 243). At the beginning of the second century BCE, the Romans peered out from the Mediterranean world onto the world of the continent. Faced with the fleeting vastness of the Po Valley, immense compared to the narrow intermountain basins they were used to, they were forced to change the scale of their plans (Chevallier 1980: 141–3). To adapt themselves to the new and spread-out dimensions, their techniques of colonization, based on town planning, underwent an analogous and almost mechanical extension. A region was born as though it were a single city and vice versa, with a super-*decumanus* along the almost 260-kilometre stretch and multiple pivots at the river valleys descending towards the Po. It was at the intersection of these courses of water with the via Emilia that centres of inhabitation developed, at distances of 10 to 25 kilometres from one another, which equals at most a two-day route for transporting goods. On the basis of the orthogonal nature of the road structure, Vidal (1922: 295) traces a parallel between the American city and those of the Roman Empire. But this parallel runs the risk of hiding a profound difference, especially evident in the case of the Emilian corridor: the organic unity which Rome's colonial system establishes between city and countryside and which makes the former the product of the latter, a sort of outgrowth charged with guaranteeing its vitality through exchange with the outside, at the intersection of local and regional circuits with continental circuits. In the Emilian countryside, they produced; in the cities, none of which arose in relation to the exploitation of a particular natural resource, they carried out tasks relating to the establishment of short- and long-range services: military, administrative, cultural, religious as much as commercial, and more generally related to the circulation of people and news. The best luck shone upon the locations provided with a structured variety of roles, mostly tertiary, which Rome, careful not to let any centre grow too large compared with the others, modulated through the regulation of flows going in their direction (Chevallier 1980: 247).

From this mechanism of controlling the city-region, imposed on the programmatic decentralization of the productive functions, derived two principal characteristics. The first was the lack of consistency of the city centres, none of which prevailed decisively over the others: from here, we derive the name Mesopolis, which means groups of medium-sized cities (Farinelli 1984: 16, 50). The second was their naturally 'transactional' function, as today we describe using a term which is erroneously believed to be applicable to the most recent forms of metropolitan growth (Gottmann 1982; Corey 1982), and which instead perfectly defines the original character of the Emilian city. Emilia Romagna is the only region of the world which takes its name from a road, because it was precisely the road, more than Rome, which was its metropolis, the true mother of its cities. When, following the process of national unification, Emilia too was inserted into the new railway area, the railways had to adapt to the pre-existing road route, like a copy with respect to the original.

88. *Urban Self-Organization and the Birth of the Quarternary*:
   The Medieval Period

In other terms: the validity, in terms of functionality, of the indexical relation between street and houses lasts, in the case in question, for 2,000 years, until after the first half of the twentieth century, also resisting in this way the beginning of the telecommunication era (Farinelli 1984: 47–84). On the other hand, despite every technological advance and despite economic institutions, the development of an urban system is still based on the same mechanism which was functional in the pre-telegraphic era, on the continual and repeated interaction of three components: economic relations which already exist at the local level and inter-urban level; the configurations which regulate inter-urban circulation of specialized information from the spatial point of view; the extension of existing relations at the local level and between cities, and the establishment of new relations (Pred 1977: 174).

But how did the Emilian corridor react to the decline of the Roman Empire, to the disappearance of the flows sent from its external capital, between the third and fourth centuries after Christ? If we limit this to the essential, in the last millennium and a half, its history consists in the attempt to make the passage from an element in a system to full, and therefore heteronomous, control, within a self-organizing system, that is, into a system capable of transforming its concrete structure through its own growing complexification, without, however, changing the logic of its organization and thus its identity (Varela 1983; Livet 1983). Self-organization is fed by disorder which knows how to transform itself into order. In order that it is effective and not a simple adaptation to the changing of the conditions which regulate the life of the system (in this case, the disappearance of the centre of command), one condition is absolutely necessary: that the urban axis be able to draw lessons from the disruption to close in on itself in a different way, generating new roles and activities capable of maintaining and re-invigorating its original functions, so as to preserve its own constitutional identity. It is thus necessary, above all, to have a material 'operational closure' (Dupuy 1982: 231–2) for which, in this case, there is undoubted evidence in recorded topographic developments, immediately before and after the first millennium, from the border cities of via Emilia: along its stretch, Piacenza expands towards the east and Rimini towards the west, as though, emerging out of the high-medieval storm, they turned back towards one another (Farinelli 1984: 24–5, 28). What, however, is vital is the fact that for any organism, the mechanisms of self-organization are the mechanisms of cognitive activity (Maturana and Varela 1980): only those which allow, alongside self-development, the birth of new functions capable of guaranteeing survival, and even progress, through the recognition and overcoming of crisis. What was the invention of the 'Studio' in Bologna at the wake of the millennium, or of the university, if not the manifestation of this activity on the part of the Bologna city-organism?

In the life of cities, the function of the university equals a financial function of a higher order, from which originate all other higher roles, the *quarternary* roles of interpretation, analysis, recycling and renewal of information (Gottmann 1976a; Gottmann 1976b: 35; Gottmann 1978: 29). That is to say, the growth of the centres along the via Emilia subverts and almost flips the dominant model which is still used in geography to explain the modalities of urban development in economically advanced countries: a model based, above all, on the coincidence between urban and industrial development (Lampard 1955). Instead, it was through the setting-up of cultural functions of a higher rank that Bologna established its leadership along the corridor in the medieval period, even if it never managed to eliminate the other structural characteristic of the Mesopolis which is reflected in its name. The term *mesos* in Herodotus (*Histories*, III.142) expresses, more than the median condition, the notion of centrality, of isonomy (§ 76). That is, the absence of domination: not even the arrival of the railways managed to impose a true relationship of domination of Bologna over the other cities in the Emilian corridor, or a marked-out role of hierarchical control. The mother of the cities remained the via Emilia, mother of the railroad just as, later, it will be the mother of the motorway which is also the copy of its route.

89. *Megalopolis: From the Map to the Metropolis*

Instead, when the metropolises of the United States began to develop predominantly tertiary functions in the course of the 1950s, revolution was spoken of. This was the revolution of the 'white collars' against the 'blue collars' of the workers and it was felt necessary to find a new name for these cities: megalopolises, a suggestive coinage and one destined to an illustrious fate (Mumford 1961; It. trans.: 243). The megalopolis was the name Jean Gottmann assigned to the concentration of cities which extends along the north-east Atlantic coast from Boston to Philadelphia (1961; It. trans.: 263), and is still described as an 'urban region' in dictionaries of geography (George 1970: 271). In other words, the mega-lopolis would be the result of the reciprocal integration of several metropolitan areas (Gottmann 1961; It. trans.: 263). It would be, in short, an 'urban system' (George 1978: 143). But is it actually like this? Does the megalopolis really exist?

At the root of the concept of the megalopolis (Gottmann 1957: 189), we have the image Aristotle gives of Babylon in *Politics*: an 'urban for-mation which has the dimensions of a nation rather than a city' (III.3, 1276a, 25). But Babylon, Aristotle soon adds, had been taken for two whole days, and one whole part of the city had not even realized. Babylon, in short, was something which seemed unified only because it was surrounded by walls, only because it appeared as such from the topographic point of view. But in reality, it did not at all act in a unitary fashion, as Herodotus also testified (*Histories*, I.191). The same seems true of the North American megalopolis.

Gottmann's confidence in the effective, unitary functioning—and on the existence of a single American super-megalopolis which included Boston, New York, Philadelphia, Baltimore and Washington among others—was based on only one concrete fact, related to communication. According to Gottmann, the exchange of telephonic messages was much more intense, absolutely and in numbers per capita, among the great centres just mentioned, than between these same cities and the other cities of the United States. But elaborating the same facts in another way, Hans Blumenfeld (1979) demonstrated that the centres in question had relatively much greater exchange with the centres of their own region and own state than with the other centres of the megalopolis. For Blumenfeld, moreover, there was no greater movement of goods and people in the megalopolitan environment compared with the metro-polis. No phenomenon of clear overlap or fusion between the cities of the megalopolis was evident. Within it, finally, with the exception of a chain of bakeries, we cannot trace any institution or company which was not present also outside it, or which provided goods and services to all the parts of the megalopolis, but to no area outside it. Blumenfeld's conclusion is decisive: if the term megalopolis signifies an aggregate of metropolitan areas, it denotes a fact; but if it means, as in effect Gottmann wished, a new and more integrated unit of settlement possessing higher and more complex characteristics with respect to the metropolitan level, it denotes instead a fiction. The goal of that fiction is to justify, through the mirage of a greater dimension of control, the abandonment of the analysis of the metropolitan level which for Blumenfeld remains the only concrete level of urban operation.

In this way, Gottmann too fell victim to what could be called the topographic axiom, unconscious reflection of the 'graphic prejudice' (§§ 62–3), according to which the geographic object is nothing other than the object which results from the totality of topographically distinguishable features. From this alleged, preliminary topographic individuality (i.e. individuality as visible on maps), Gottmann inferred the birth of a new urban entity, the specificity of the megalopolis in essential terms. Gottmann made the uniqueness of the dimensions of the agglomerate (its topographic primacy and the exceptionality of its form) correspond to the uniqueness of the internal mechanism, and its functional primacy (§ 64)—without thereby profiting from the lesson of Babylon which it recalled before all else.

90. *Urban Systems*

In what sense, then, can we speak of urban systems? Is it possible to distinguish these today?

In trying to respond, we must immediately make two clarifications. The first concerns the environment within which, in the technical sense, the expression has been affirmed in the last quarter-century: the regional or national level. An urban system is a group of cities interconnected within a state territory or one of its parts (Berry 1964; Pred 1977; Bourne and Simmons 1978). The second clarification concerns the nature of settlements and consists in the disappearance of all information related to the material constitution of the city, to its physical character or to its concrete space from the urban system. This is an authentic revolution, definitively completed in the middle of the 1970s (§ 42).

Its origin is perhaps located in the idea of the 'urban field' (Friedman and Miller 1965; Friedman 1973), understood as the basic territorial unit of the post-industrial city, the city precisely dedicated more than any other to tertiary and *quarternary* activities (production of services and specialized information). This is a vast area distinguished from the traditional city in two ways. First of all, it cannot be visualized as a totality but, owing to its dimensions, can be carried out only one part after the other—thus, in sequence. Further, it is identified not from the continuity of the urban fabric, as would be the case of the urban region, but from the use which people make of their environment, to the point that its external limits coincide with spaces periodically used by its inhabitants for recreational purposes. At first sight, this area differs slightly, with its inclusion in the conurbation (§ 87) of rural zones within the urban environment, theorized on the eve of the First World War to overcome 'the traditional idea of countryside and city on which we were raised' (Geddes 1915; It. trans.: 60). But one difference exists: the functional and no-longer-topographical distinction of limits, the same which begins the destruction of the concept itself of the urban limit, at least in its material version. Up until then, the extension of the city had remained ideally Euclidean (§ 4). The removal of the static and geometric definition of its border involves the crisis of topographic homogeneity and continuity of the city, and only the isotropic character of its parts remains, at the start, unchanged.

We enter in this way into the recognition of urban or urbanized space as a space functionally distinguished by flows of people, money and goods, which on the plane of models involves a radical change in which the temporal dimension takes precedence over the spatial one. The investigation is concentrated on the confrontations among a group of individual behaviours related to the economies of agglomeration, to the reaction speed of the population, to the propensity to segregation, to the sensitivity to distance or congestion and so on. These behaviours are specified a priori by a series of quantitative parameters, that is, by schemas whose formalization no longer maintains a relation with urban form itself. In reality, it is these parameters, and not cities, which host the actors they refer to, so that they are in this way systematically linked (Pumain et al. 1989; Beguin 1991).

It is not perhaps a coincidence that this revolution, this authentic dematerialization of the urban, this sudden evacuation of the physical city from the plane of analysis intervenes precisely in conjunction with the global decline of the real-estate market at the beginning of the 1970s (§ 32). According to David Harvey (1985: 6–7), this event opens a new phase in the process of urbanization of capital. In any case, the evident paradox consists in the fact that exactly in the moment of its maximum expansion, often following a lull which at least in Italy is age-old (Carozzi and Rozzi 1980), and coinciding with the maximum momentum for its construction, the city enters into crisis regarding its original task: to elaborate an image of itself in which it can be recognized and succeed in thus controlling its own development. It is precisely this paradox, in which the city ceases to properly be the city, which is the last sign of the breaking of the indexical link between streets and buildings: the only systematic thing that has happened in the urban field in the course of the twentieth century.

91. *The Fordist City*

Starting from the end of the First World War and up to the 1980s, different hypotheses or generations of spatial models related to the use of the land within cities were adopted and then abandoned or modified. All were thought able to respond to the question: Is it possible to find, in the internal disposition of urban functions, some form of order whose recurrence implies the resemblance of nature and the devices of the development of the cities themselves?

The first response was Burgess' 'Concentric zone model' (1925), made up of a centre and five rings, each featuring a specific task. The central core plays the role of command of civic, economic and social life. Around it extends a transitional zone, a ring in the process of downgrading in which commercial activities and small industry are mixed with residential activities, surrounded in turn by the compact halo of workers' habitations. On the back of this, the rings of habitations of the middle and upper classes are developed; and outside the edges of the city proper extends the ring of small city-dormitories inhabited by commuters. Decades later, it was easy to accuse this model of excessive rigidity (Beaujeu-Garnier and Chabot 1963; It. trans.: 360ff), and precisely through the explanation of the development of the city of Chicago, which was Burgess' objective. At the same time, the attempts towards the implementation of this model in other contexts signalled the influence of different factors which were absent in it, starting from the weight of historical inertia and the duration of certain uses related to the fixed nature of the capital invested in cities (Racine 1971: 400–02). None of these critiques, even if correct, catches the fault nestled deep in the origins of the model itself, a fault also present in the second schema born from the research of the Chicago school of sociology: that suggested by Roderick Duncan McKenzie (1933) and taken up and clarified in detail by two geographers, Chauncey D. Harris and Edward L. Ullmann (1945), under the name of the 'Multiple-Nuclei Model'. In it, the use of urban land is no longer divided around a single heart but a plurality of ganglions, diverging from one another but each possessing a power of attraction, produced by the internal specialization of functions. The complete destruction of the circular order results from this, and a kind of patchwork replaces the series of concentric rings.

Antonio Gramsci explained that Fordism, the practice of mass production for mass consumption introduced in 1914 by Henry Ford's

automobile magnate, is based on the inclusion of the city, and in particular its system of transport, within production itself (1975[1948–51]:
2145–6). It is on transport itself that the third model is based, that of
'radiating sectors', formalized as the 'Sector Model' by the economist
Homer Hoyt (1939). Its originality is based on the recognition of the tendency towards the axial development of urban areas, that is, along the main
routes of communication, so as to create a system in the shape of a star.
According to Hoyt, the growth of cities along a particular artery is often
based on an identical type of use of the land and on the upsurge of a single
activity. In this way, the radial street axes originating in the centre act as
the main element distinguishing functional urban zones which would
assume the form of circular sectors. Even Hoyt's model, like the previous
ones, is the result of a generalization featuring two characteristics: it is
born from empirical research in an American metropolis; and it aims at a
comprehensive reconstruction of the city unit focused on the bijective
correspondence between an individual area and a single function.

This is indeed the point, because this is exactly the same correspondence between a single unit of labour power and a single function which
the Fordist model of the assembly line is built on (Marazzi 1994: 17).
First of all, however, this correspondence condenses cartographic logic,
the logic of which each of the above sections, from first to last, has exemplified a different version or aspect. Urban models developed in the
interwar period are already topographic: the city already imitates the
map, since the city's functioning already obeys the logic of the map.

92. *The Keynesian City*

But this is the last time. For the Chicago school, the city was a constel-
lation of natural areas, each with its characteristic environment and each
with its specific function to perform in the urban economy in its totality.
Natural area here means, from the sociological viewpoint, a zone in
which individual types tend to be concentrated who, because of one trait

or other, consider themselves to be homogeneous. Without this postulate, scientific observation of social reality would itself be impossible (Pizzorno 1967: *xvii–xviii*). But the homogeneity in question is only the reflection, in turn, of the homogeneity connected to Euclidean extension which we have discussed numerous times so far, which descends directly, even if unconsciously, from the features of the map (§§ 4, 90). Its validity begins to fade with the passage from the Fordist city to the Keynesian city.

The Fordist city is the city of production; the Keynesian city, the city of consumption. It takes its name from John Maynard Keynes, the English economist who theorized, in the 1930s, the intervention of the state into the gestation of fiscal and monetary policy intended to incentivize urbanization from the side of demand and to thus resolve the problem of unemployment. It would have been difficult for capitalism to survive in the postwar period without state-promoted consumption, financed by debt (David Harvey 1985: 206–07). It involved the restructuring of the territory and was translated into the immoderate growth of the peripheries: a way to make the products of construction firms, petrol and automobile companies, and rubber factories necessary, which transformed the city into a giant artefact for income redistribution. Urban boundaries changed and the agglomerates were spread out following the style that in Europe, and especially in Italy, involved low population density far less than in the United States. In Europe, functional regions (in economic and urban terms) sprang up, centred, especially in the north, on the capacity of a large city for radiating outwards, a large city capable of acting as a coordinating centre in relation to a network of medium and small-sized locations surrounding it (Gambi 1972: 55–8).

But beyond differences from country to country which expressed the weight of different historical inheritances, it was a process that continued through the 1970s across the world, more and more clearly directed by logic and the interests of big business. These began to break free totally from the frame of reference of the nation-state and to exploit across the board, according to the logic of the universality of abstract labour, different relationships in the new articulation of the world economy: national, international (between two individual states), multinational (between more than two states) and global (Beaud 1987). We can thus understand the birth of the concept of 'urban field', a faithful portrait of a development oriented around distinctive fields of consumption, and on the dissolution of the rigidity of every previous boundary. At the same time, the abstraction of finance capital, which the Keynesian city obeys, explains the mathematical abstraction of models that have governed its analysis. If at the beginning it keeps, in part, at least the isotropic quality (§ 90), in the 1970s it definitively and in functional terms loses this, because its body is less and less the expression of decisions taken within it and more and more the result of external choices: for example, the development of a branch of a bank or the installation of an advertising agency, created by the parent companies on which they depend, which are multinational and towards which, like urban components, they are oriented.

Contrary to the Fordist city—national, topographic and visible—the Keynesian city begins to be transformed into a transnational organism irreducible to the topographic model and invisible in its mechanisms. This is because in 1969, the same year as the first moon landing, the first network of electronic communications was born in the United States (Gillies and Cailliau 2000), and the matter which surrounds us begins to be transformed into immaterial units of information—an event which can only be compared to the mythical confrontation between Ulysses and Polyphemus (§§ 0, 50, 58, 59). Then, space had come to light for the first time; in the 1970s, taken into the network, it began to die.

93. *The Informational City*

The informational city prepared throughout the 1970s and 80s no longer functionally resides within a national group, even less a regional one. It is even less able than the Keynesian city to be reduced to the simple topographic model, and is only half-visible. Starting from the beginning of the 1970s, the logic of the Keynesian city enters into crisis, at least in the United States (§ 32), and the urban process returns to being dominated by questions related to the organization not of consumption but, rather, of production (David Harvey 1985: 212–21). This did not at all signify the return to the topographic form as a principle of definition of the urban, or to attention towards expressions of construction as a principle of the reconstruction of generative phenomena of the city. On the contrary. The reason for the growing abstraction of analytic models with respect to the visible fact is mimetic. It depends on the fact that it is the same process of production, including that of the city, that is progressively engulfed by the immaterial and the invisible, by the information revolution. This expression signifies the increasing grip exercised over the functioning of the world by the system of electronic flows, which determines the field of the information economy and, thus, impetuously redesigns the whole face of the Earth.

The arrival of this system goes hand in hand with a series of concomitant phenomena, linked to it so intimately as to barely allow us to distinguish between cause and effect: (a) the abandonment of the Keynesian programme of social redistribution on the part of the state; (b) the acceleration of the internationalization of economic processes; (c) the decentralization of production linked to the flexible location of a plant; (d) the development of new technologies based on the treatment of information. The result of these upheavals is condensed into the differentiation of labour in two equally dynamic sectors: the formal economy based on information and the informal economy, meaning the illegal economy, which exploits every degraded kind of labour (Castells 1989: 13, 25–7, 127–37, 225). The half-invisible character of this new urban form derives from the illegality of the informal economy and, simultaneously, the electronic rather than material nature of flows of information. The logic of the system of flows disarticulates, socially before spatially, all local structure. At the same time, local values remain indispensable, precisely because the information economy rests upon them. And precisely because its raw material is information, it is based on the immaterial capacity for symbolic manipulation, which is another way of saying 'culture': something, thus, which depends on the overall setting in which we are educated, which varies from place to place and which is the main resource labour power offers inside the global market (ibid.: 351). When, for example, it is repeated that immigrants arrive in Europe to do jobs that the locals no longer want to do, something else is actually meant: it affirms that some attitudes and mental habits (think of elderly care) have disappeared in Europe, despite their need still being experienced in social terms. For this reason, it is necessary to turn to people who, coming from places other than the local area, still retain such cultural dispositions.

In the course of the 1980s, in short, the growth of the system of electronic flows translated into the definitive crisis of Euclidean-topographic space, reconfigured the relation between this kind of space and the plurality of places, between the model of space and that of place. This reconfiguration, like all relations, presupposed the possibility of a distinction between one and the other, the existence of a reciprocal gap produced by the difference in their nature, just as it had been for all of modernity. In this way, the functioning of the world could still be thought in terms of a dialectical articulation between environments that are different but integrated, lying between two senses (§ 13) that both refer to the same signified—just as the informational city consists of a double model, composed of two different social and topographic segments, each animated by its own logic.

94. *A Whole Turn*

Today it is no longer like this, or at least, less and less like this.

We return to Ptolemy and to his advice of assuming the map and not the globe as the model of the world (§ 48), peremptory at the point of resembling a true prohibition, so much so that when in 1492 the

Monacan Martin Behaim constructed the first modern globe of the Earth, he called it 'the apple'—just like the fruit of the tree of knowledge of good and evil, the fruit of original sin. For men in the medieval period, God was a globe, a sphere that had its centre everywhere and its circumference nowhere. In the golden period of the construction of globes, that is, in the seventeenth century, Pascal will apply this definition to nature. The globe, in fact, is the prototype of Baroque sculpture, a work conceived for a spectator who circles it. But if it is the statue of the Earth, where does the spectator who remains in this way external to the Earth walk, if not on something imagined infinite and void in contrast with the limitedness and fullness of the globe which those who look are excluded from? And how is it possible to imagine any of this as being real before the first astronaut, that is, the first truly postmodern subject, the first to be removed, with his circling the Earth, from the modern equivalence (§ 5) between world and cartographic image of the world?

To circle the globe, as Ptolemy advises against, does not only lead to the implicit admission of the existence of absolute void, and not only produces the *horror vacui* which paralysed Western culture up to the end of the eighteenth century, but it also corresponds to a specific practice: a practice in which knowledge is the result of a process implying time— literally, the outcome over a period of a circuit around something, which does not finally lead anywhere if not back to the starting point. To stay, instead, still, and slide the globe around with one's hand, which is the sole alternative, involves instead the idea that knowledge is based on sight and touch in the same way, and not on their separation. It is, however, telling that in both cases (if Ptolemy were paid no heed), the fundamental rules of modern epistemology would be contradicted, which are rigid protocols for the relationship between subject and cartographic representation fixed on perspectival vision: the subject remains still and, in order to know, only needs the gaze which travels the distance to the object instantly, meaning atemporally (§§ 4, 5, 9, 34, 72). That is to say, if one uses the globe, it is impossible to be Kantian and impossible to distinguish time from space. For this very reason, Nietzsche sustained the necessity of dancing-thinking, that is, thinking in such a way as not to separate the two. Space does not exist for the subject circling the globe, neither where the subject is located nor on the sphere: the steps of the subject obey a measure which is certainly not standard, and there is no scale on the globe. Instead, the proportions of the globe's parts depend on reciprocal internal relations, according to an exclusively self-referential logic (§§ 0, 8). According to the logic of the globe, on which there are no straight lines (§ 9), a large volume corresponds to a small surface (Volk 1995: 10–13). This principle corresponds to the principle of capitalist accumulation which, if in the past was served by space (§ 4), today appears more and more modelled on its opposite: selective and thus discontinuous, fragmentary and thus non-homogeneous, an-isotropic since straight in its absence of a centre, and based conversely on the existence of a plurality of possible virtual centres, just as occurs on the surface of the globe (§ 8).

In exactly the same way, the city is selective and thus discontinuous, fragmentary and thus non-homogeneous and not at all isotropic. This is the city which has made its appearance on the Earth in the last decade and already controls it: the global city, within whose mechanism space and time can now no longer explain almost anything, and topographic appearance, the visible, is just an outer skin from which nothing plausible or concrete can be obtained about the functioning of the world (§ 24). This is the city, which we all inhabit even when we have the impression of being far away from it or being on the other side of the planet— that in which, as in the hotels of Hong Kong, clocks signal the time in nanoseconds, a measure totally useless for any appointment.

95. *The Global City*

After the aeroplane attack on Manhattan's Twin Towers on the 11 September 2001, we have come to know something more about the geography of the skyscraper (§ 34). In December 2002, the number of victims of the destruction of the World Trade Center was fixed at 2,792. Just under two-thirds came from the state of New York and just under a quarter from the adjacent state of New Jersey, only around 100 from the rest of the United States and all the others, more than 300, from the rest of the word, especially from Asia. In urban life, almost all relations begin with exchanges between strangers (Meier 1962; It. trans.: 78) but never in the history of humanity have these relations been, at long range, so intense. The tragic consequences of the attack on Manhattan give another proof, if there was need of it, of the insignificance of physical proximity for the functioning of the world. It certified the end of the megalopolis, if it ever truly existed (§ 89). The number of citizens of the states which make up the rest is insignificant, except the 60 victims from Connecticut: more than double the number of people who were buried in the collapse came from Bangladesh than from Pennsylvania. This dismal cartography properly illustrates the nature of the cities that command the global economy today, the principal seats of financial activity and the innovation related to it: global cities, as they have been called since the beginning of the 1990s (King 1990).

Global cities are not necessarily the largest cities on the Earth. When the first of them were recognized as such, according to the classification of the United Nations, the cities that surpassed 10 million inhabitants numbered 13: Tokyo with 25 million, Sao Paolo with 20, New York and Mexico City with 15, and then Shanghai, Bombay, Los Angeles, Buenos Aires, Seoul, Beijing, Rio de Janeiro, Calcutta and Osaka. Of them only the first four are at this point counted among the global cities, to which are added London, Paris, Frankfurt, Zurich, Amsterdam, Sydney, Hong Kong (Sassen 1994: 13–14). As we can see from a simple and immediate comparison, the European and North American cities control, the Asian and South American cities expand dramatically while African ones do not even appear. Although the list of global cities is to be understood in a dynamic way, just as that of the mega-cities, changes happen slowly within it, also caused by the particular logic of financial industry, typically hierarchical and selective: while its activities extend over the base continually, concentration grows at the top. This occurs because global cities are integrated by a true financial production line: towards the middle of the 1980s, for example, Tokyo was the main exporter of the primary material called coin; New York was the great centre of manufacturing, based on the transformation from crude or credit forms of money into new products meant to maximize their return; London, strong from the administrative heredity of the old British Empire, was the meeting point of minor financial markets spread out over the whole world (ibid.; It. trans.: 34, 67).

But these functions depend mainly on the information revolution of the system of communications and on its transformation into an electronic network. Consequently, there is no relationship between the lack of labour power employed and the enormous scale of financial transactions that pass through the nodes of the network (Abu-Lughod 1999: 327–8). In other terms, global cities, like mega-cities, 'are connected globally and disconnected locally, physically and socially' (Castells 1996; It. trans.: 466), meaning from the point of view of the Euclidean city, the city born from the table and which ended in the crisis of the Fordist city (§§ 4, 24, 61, 63, 72–3, 76–9, 81, 91). Such cities come into existence and develop, for the most part, in an invisible space of electronic flows. Precisely because they are composed, like the rest of the national territories, of fragments of what previously tended to be visible, continuous, homogeneous and isotropic space, they host a socially fragmented and culturally non-homogeneous population—as was sharply shown to the world by the murderous cartography on living flesh of 11 September 2001.

96. *The Electronic City*

Between the Euclidean city and the electronic city, the same relationship exists as between analogue reality and digital reality. The second is not the *opposite* of the first but its *development*: just as digital technology, which consists in the treating of information in discrete form (that is, discontinuous) according to binary descends from the logic that governs the relation of the sign to the table, where the latter can only be or not be. Only, however, this is a development which seems to erode all that exists, despite deriving from it. The electronic city is problematic for its inhabitants because it reintroduces the invisible, re-enchanting the world in functional form, which, at least in Western urban terms, had already been disenchanted (§§ 72–3). Further, it creates pockets of unprecedented wealth for the few—those that Manuel Castells calls 'globopolitans' (1997; It. trans.: 76), the high functionaries of the network who are half-living beings and half-flows—and increasing poverty for everyone else (Kotkin 2000: 182). The instruments of human interaction, production and consumption are miniaturized, dematerialized and unhooked from any fixed location. An email address, in fact, equals the proper name of an individual, signalling the simultaneous redefinition of space, personal identity and subjectivity which is emerging from the growth of the network (Mitchell 1995: 8). In this sense, cartographic logic in which only proper names exist (§§ 17, 82) is totally overthrown. Together with it, the decisive presupposition of any possibility of knowledge seems to disappear, the faith that there exists a relationship between what we see and the functioning of the world (§ 24).

The best definition of all of this, of the new world, remains the least recent, coined in the first half of the 1960s by Melvin M. Webber (1964: 7–37), who already spoke of 'nonplace urban realm'. With this expression, Webber prefigured something that subsequently actually happened: the assigning of higher functions, traditionally connected to the city, in places completely devoid not only of any connotation of the city but also of the local, impossible to locate exactly, even in terms of place. It is this which, for example, happened in New Jersey in the 1990s, just outside New York, along the rural corridor called Secaucus, known until then only for its pig farms. Today, however, it is the United States' capital of electronic money—and since 1998 it has moved more money than

New York City itself—the capital of capitalism, because all the operations which pass through the automatic counters that we call cash machines are regulated along it (Abu-Lughod 1999: 530–1). Or think of the most recent urban American frontier, the edge city, popping up around the old large centres in California, on the Atlantic coast, in Arizona, in Texas, and different from all the cities up to this point appearing on the Earth. This is any place that obeys the following characteristics: (a) it has at least 3 million square metres rentable as office space (the workplaces of the information age); (b) it has more than a tenth of its surface rentable as retail space; (c) it has more jobs than it does bedrooms; (d) it is perceived as one unit by those who inhabit it; (e) it is linked not by trains or metropolitan lines but by motorways, flight routes and satellite dishes (Garreau 1991: 4, 6–7), and, above all, by information networks. Such peripheral areas owe their form and their nature to these factors as well as their possibility of existence. This leads us, in the end, not to think that we are actually about to bid yesterday's world goodbye, to vault into an absolutely new and unknown world.

On the contrary.

An information network presupposes a street, even a street network, which serves as a guide for the installation of underground cables necessary for its power supply. But once the cables are installed, the relation between what is above and what is below is overturned, because the former begins to depend on the latter which only a moment before was not there, and which, however, now reduces it to its reflection. To sum up: there is no centre and, thus, there is no space; individual identity is threatened, and what we see is no longer enough to orient ourselves. This is the present condition—but it is also the archaic one.

97. *The Impossible Labyrinth*

Have we ever truly understood what a labyrinth is? At the beginning of this volume, we supplied a possible explanation (§ 8). Henri Focillon

(1943; It. trans.: 35) wrote regarding architecture that its most wonderful aspect consisted in having conceived of and created its own universe. Giving a definite form to what is hollow, it had produced something 'inverse' to what was outside, to the external dimension which initially was the only dimension given to man. This production remained thereby foreign to all things. The submission of verticality to horizontality at the origin of Greek culture was an analogous revolution in importance and reach. This was the cancellation of verticality in horizontality, which is expressed in the figure of the labyrinth.

A device emerges from this cataclysm which seeks to 'grab nature as a totality organized by calculable forces' in order to 'develop it for use' and whose mode is 'exactness of representation' (Heidegger 1954; It. trans.: 14, 15, 16). This is the table, the map, the chart from which, long before modernity, all technique emerges. In it, the first of the laws coined by Melvin Kranzberg already applies (1985: 50), laws for the techno-logical revolution which today permeate the production and circulation of information. This law maintains that technology is neither good nor evil but is not neutral either. The first proof of the non-neutral character of the map consists in the demonstration of its formidable ontological power, capable of deciding on the existence or non-existence of things— just as the very figure of the labyrinth serves to testify. This was signalled in the first pages of this volume: representing the labyrinth is impossible, because every representation implies, in its material objectivity, in its concrete status as thing, a centre, or, in any case, a definition of a system of centres. On the contrary, the labyrinth corresponds by definition to the absence of centrality. The design (like in sculpture) of the labyrinth does not exist, because if the labyrinth is sculpted or drawn, then it is reduced to a table, ceasing automatically to be a labyrinth, since thereby it is inevitably found to possess a centre. In this way, the term indicates an archetypal contradiction between what can be thought, between existential situations which occur to us (who has never lost their way, even once?) and what can be represented: between what exists and what *subsists*. The labyrinth denounces the impotence of the system of *trapesography* ('table-graphy') to translate every condition into a design and every situation into a schema. At the same time, the fear shown towards it reveals the fact that our entire world has been built on its opposite—that is, on the table.

In this sense, modernity invents nothing in relation to the original movement but continues along in an already established direction. Writing, as we recognize it today, was born in Mesopotamia more than 3,000 years ago from the reproduction of the things of the world on a table, through a more or less schematic sketch: a pictograph (Bottéro and Kramer 1990; It. trans.: 33). The pictograph thus becomes the interpretative model for natural occurrences, in the sense that phenomena held to be heralds came to be considered as a sort of divine pictograms: things able to anticipate what would later have occurred. In this way, the entire sublunary sphere was considered the support for divine scripture (Bottéro 1987; It. trans.: 136–41). On one side there was the language table of the gods, on the other side the world remained as it was, although interpreted on the basis of the former. Only the assimilation of the world to a plane, which the Greek labyrinth is both agent and trace of, transformed knowledge into a continuous function between two tables: the table of alphabetical or geometrical writing and the table of the world. This is the passage which immediately precedes mapping (§ 37).

The Greek labyrinth, the original representation which as such cannot be represented, the original writing which as such cannot be written, reflects the triumph of the horizontal dimension over the vertical structure of the world. Another labyrinth, less well known, is entrusted the task of conserving the memory of the true nature of our planet.

98. *Critique of Cartographic Reason*

Take a pencil (which Ulysses sticks into Polyphemus' eye) and trace a straight line (which Ulysses orders the olivewood stake to be reduced to) on a blank page (which Hobbes reduces the world to). It is thus that the Earth is born, the Earth we have in our head, the only Earth that counts, and geography is born. If above the line, you trace a small *s*, and underneath a large *S*, you will have the sign which Ferdinand de Saussure (1916; It. trans.: 158–62) used to designate the sign of the sign: the relation between the signifier and the Signified and the result of the terrible history, remembered at the beginning of this volume, of the dismembering of Dionysus at the hands of the Titans and his successive recomposition by Apollo. Incidentally, Apelles, who, according to the famous nursery rhyme, is Apollo's son, must have learnt something from his father about the technique of recomposing a body cut into pieces and torn apart in a spherical form.

But if in place of the small and large *s* you draw a small and large *f*, you will have an even older model, almost primordial: the model of the Egyptian labyrinth described by Herodotus (*Histories*, II.148), which surpasses everything speakable—even more so than the pyramids. It is a construction larger than any other human work. At first sight, it is a very different building to the Greek labyrinth, but in reality it is the same. What changes, however, is the most important thing: the process which leads to its form. This is despite the fact that in appearance it is identical to the house of the Minotaur. The Egyptian labyrinth is perfectly geometrical: it is composed of twelve adjoining covered courtyards, arranged six on the side to the north and six on the other side to the south, so that the door of each opens exactly in front of the door of another, and an external wall, made of stone like the rest, surrounds all of them. Each courtyard is surrounded by a white colonnade which enters into a series of rooms, 1,500 in total, and each courtyard and each room is so full of things to see that it is impossible to cross it in a straight line—just as in the luxury streets reserved for shopping one proceeds zig-zagging from one shop window to another.

But if everything is geometrical, oriented and symmetrical, what makes up labyrinthic nature overall, and where is the labyrinth? The answer is in the relation between what is above and on the surface, and what is underground. The visible structure is in fact the reflection, the faithful copy of what is underneath, invisible and inaccessible, watched over by priests, which guards the tombs of the 12 kings whom the construction owes itself to and the tombs of the sacred crocodiles. Often in *Histories*, Herodotus limits himself to the description of the exterior of buildings. In the case in question, he instead forcefully underlines having also seen what is inside. What comes to mind is Humboldt's pride in claiming to have also crossed the interior of the American continent, different from the previous travellers who had limited themselves to skirting its coast. But for Herodotus, the first to describe the *ecumene* in spatial terms (§ 79), what is below—which explains all the rest just as the subterranean information networks today explain the global city— remains inevitably precluded and expunged from the vision and account of the world. For this reason, he is the father of geography, even as such his history begins with Egypt and the story of Egypt finishes with the labyrinth.

Today, we must start again from this history, since it is untrue that the postmodern epoch—as we are used to calling our own—is determined by the 'precession of simulacra', by the precedence of the map over the territory (Baudrillard 1981: 10). This was certainly true throughout the whole of the modern period and it was already true for Anaximander. On the contrary, our world is based precisely on the end of this anticipation, because by now the map and the territory are no longer distinguishable from each other, in the sense that what we see of the latter has completely assumed the form and nature of the former. Thus, we manage to understand little of how the world functions. For this very reason, we must return to discover the labyrinthic character not of the Earth's surface but of our planet, what is above it and what is below it, *Gé* as well as *Cton*, which, even if we have too long forgotten, are a single thing—something the story of the labyrinth of the 12 kings and sacred crocodiles still recounts, to whoever wishes to listen.

Works Cited

ABU-LUGHOD, Janet. 1999. *New York, Chicago, Los Angeles: America's Global Cities*. Minneapolis: University of Minnesota Press.

AMMERMANN, Albert J., and Luigi Luca Cavalli-Sforza. 1973. 'A Population Model for the Diffusion of Early Farming in Europe' in Colin Renfrew (ed.), *The Explanation of Culture Change: Models in Prehistory*. London: Duckworth, pp. 343–57.

ANATI, Emmanuel. 1960. *La civilisation du Val Camonica*. Paris: Arthaud. **Italian translation**: 1964. *La civiltà della Valcamonica*. Milan: Il Saggiatore.

APPADURAI, Arjun. 1988. 'Putting Hierarchy in Its Place'. *Cultural Anthropology* 3(1): 36–49.

ARENDT, Hannah. 1958. *The Human Condition*. Chicago: University of Chicago Press. **Italian translation**: 1989. *Vita Activa. La condizione umana*. Milan: Bompiani.

ARNHEIM, Rudolf. 1969. *Visual Thinking*. Berkeley: University of California Press. **Italian translation**: 1974. *Il pensiero visivo*. Turin: Einaudi.

AUJAC, Germaine. 1987a. 'The Growth of an Empirical Cartography in Hellenistic Greece' in J. B. Harley and David Woodward (eds), *The History of Cartography, Volume 1: Cartography in Prehistoric, Ancient, and Medieval Europe and Mediterranean*. Chicago: University of Chicago Press, pp. 148–60.

———. 1987b. 'The Foundations of Theoretical Cartography in Arcaich and Classical Greece' in J. B. Harley and David Woodward (eds), *The History of Cartography, Volume 1: Cartography in Prehistoric, Ancient, and Medieval Europe and Mediterranean*. Chicago: University of Chicago Press, pp. 137–47.

AUYANG, Sunny Y. 1995. *How Is Quantum Field Theory Possibile?* New York: Oxford University Press.

AYMARD, Maurice. 1978. 'La transizione dal feudalesimo al capitalismo' in Ruggiero Romano and Corrado Vivanti (eds), *Storia d'Italia. Annali, Volume 1: Dal feudalesimo al capitalism*. Turin: Einaudi, pp. 1131–92.

BAKHTIN, Mikhail M. 1975. *Voprosy literatury i estetiki*. Moscow: Progress Publishers. **Italian translation:** 1979. *Estetica e romanzo*. Turin: Einaudi.

BADE, Klaus J. 2000. *Europa in Bewegung. Migration vom späten 18. Jahrhundert bis zur Gegenwart*, Munich: C. H. Beck. **Italian translation:** 2001. *L'Europa in movimento. Le migrazioni dal Settecento a oggi*. Rome: Laterza.

BALODIS, M. J. 1988. 'Generalization' in R. W. Anson (ed.), *Basic Cartography for Students and Technicians*, VOL. 2. London: Elsevier, pp. 71–84.

BARBER, Peter. 2001. 'Mito, religione e conoscenza: la mappa del mondo medievale' in *Segni e sogni della Terra. Il disegno del mondo dal mito di Atlante alla geografia delle reti*. Novara: De Agostini, pp. 48–79.

BARTHES, Roland. 1954. *Michelet par lui même*. Paris: Seuil.

BARBUJANI, Guido, and Robert R. Sokal. 1990. 'Zones of Sharp Genetic Change in Europe Are Also Linguistic Boundaries'. *Proceedings of the National Academy of Sciences, USA* 87: 1816–19.

BATESON, Gregory. 1979. *Mind and Nature: A Necessary Unity*. New York: Dutton. **Italian translation:** 1984. *Mente e natura. Un'unità necessaria*. Milan: Adelphi.

BAUDRILLARD, Jean. 1981. *Simulacres et simulation*. Paris: Galilée.

BEAUD, Michel. 1987. *Le système national/mondial hiérarchisé. Une nouvelle lecture du capitalisme mondial*. Paris: La Découverte.

BEAUJEU-GARNIER, Jacqueline, and Georges Chabot. 1963. *Traité de géographie urbaine*. Paris: Colin. **Italian translation:** 1970. *Trattato di geografia urbana*. Padova: Marsilio.

BECK, Hanno. 1961. *Alexander von Humboldt, Volume 2: Vom Reisewerk zum 'Kosmos', 1804–1859*. Wiesbaden: Steiner.

BEGUIN, Hubert. 1991. 'Les modèles urbains dynamiques en perspective'. *L'espace géographique* 20: 117–18.

BELLOSI, Luciano. 1980. 'La rappresentazione dello spazio' in Giovanni Previtali (ed.), *Storia dell'arte italiana*, VOL. 4. Turin: Einaudi, pp. 3–39.

BENJAMIN, Walter. 1955. 'Das Kunstwerk im Zeitalter seiner technischen Reproduzierbarkeit' in *Schriften*, VOL. 1. Frankfurt am Main: Suhrkamp, pp. 366–405. **Italian translation:** 1966. *L'opera d'arte nell'epoca della sua riproducibilità tecnica*. Turin: Einaudi.

BERDOULAY, Vincent. 1981. *La formation de l'école française de géographie, 1870–1914*. Paris: Bibliothéque Nationale.

BERENGO, Marino. 1975. 'Le città di antico regime' in Alberto Caracciolo (ed.), *Dalla città preindustriale alla città del capitalism*. Bologna: Il Mulino, pp. 25–54.

BERMAN, Marshall. 1982. *All That Is Solid Melts into Air: The Experience of Modernity*. New York: Simon and Schuster. **Italian translation**: 1985. *Tutto ciò che è solido svanisce nell'aria. L'esperienza della modernità*. Bologna: Ll Mulino.

BERRY, Brian. 1960. 'The Quantitative Bogey-Man'. *Economic Geography* 36: 282.

———. 1964. 'Cities as Systems within Systems of Cities'. *Papers and Proceedings of the Regional Science Association* 13: 147–64.

BERTIN, Jacques. 1967. *Semiologie graphique: Les diagrammes, les réseaux, les Cartes*. Mouton: La Haye.

BIANCHETTI, Serena. 1997. 'Conoscenze geografiche e rappresentazione dell'ecumene nell'antichità greco-romana' in Claudio Tugnoli (ed.), *I contorni della Terra e del mare. La geografia tra rappresentazione e invenzione della realtà*. Bologna: Pitagora, pp. 51–92.

BIASUTTI, Renato. 1962[1947]. *Il paesaggio terrestre*. Turin: Utet.

BLOCH, Marc. 1952. *Les caractères originaux de l'histoire rurale française*. Paris: Colin. **Italian translation**: 1973. *I caratteri originali della storia rurale francese*. Turin: Einaudi.

BLUMENFELD, Hans. 1979. 'Megalopolis: Fact or Fiction?' in *Metropolis . . . and Beyond: Selected Essays* (Paul D. Spreiregen ed.). New York: Wiley and Sons, pp. 116–26.

BLUMER, Walter. 1964. 'The Oldest Known Plan of an Inhabited Site Dating from the Bronze Age, about the Middle of the Second Millennium BC'. *Imago Mundi* 18: 9–11.

BOCCHI, Francesca. 1982. 'La "terranova" da campagna a città' in Guiseppe Papagno and Amedeo Quondam (eds), *La corte e lo spazio: Ferrara estense*, VOL. 1. Rome: Bulzoni, pp. 167–92.

BOLTER, J. David. 1984. *Turing's Man: Western Culture in the Computer Age*. Chapel Hill: University of North Carolina Press. **Italian translation**: 1985. *L'uomo di Turing. La cultura occidentale nell'età del computer*. Parma: Pratiche.

BONAPACE, Umberto. 1990. 'Metodi e contenuti della cartografia' in *Appunti di didattica della geografia*. Genova: Associazione Italiana Insegnanti di Geografia, Sezione Liguria, pp. 11–20.

Borges, Jorge Luis. 1985. 'Atlante' in *Tutte le opere*, VOL. 2. Milan: Mondadori, pp. 1311–1423.

Botero, Giovanni. 1598. *Delle cause della grandezza delle città. Libri tre.* Appendix to *Della ragion di stato. Libri dieci.* Milan: Pacifico Pontio.

Bottéro, Jean. 1987. *Mesopotamie: l'écriture, la raison et les dieux.* Paris: Gallimard. **Italian translation**: 1991. *Mesopotamia. La scrittura, la mentalità, gli dei.* Turin: Einaudi.

———, and Samuel Noah Kramer. 1990. *Lorsque les dieux faisaient les hommes: mithologie mesopotamienne.* Paris: Gallimard. **Italian translation**: 1992. *Uomini e dei della Mesopotamia.* Turin: Einaudi.

Bourgeois, Émile. 1920. *Notice sur la vie et les travaux de M. Paul Vidal de La Blache.* Paris: Institut de France.

Bourne, Larry S., and James W. Simmons. 1978. 'The Nature of Urban Systems' in Larry S. Bourne and James W. Simmons (eds), *Systems of Cities: Readings on Structure, Growth, and Policy.* New York: Oxford University Press, pp. 3–15.

Braudel, Ferdinand. 1949. *La Méditerranée et le monde méditerranéen à l'époque de Philippe II*, Paris: Colin. **Italian translation**: 1953. *Civiltà e imperi del Mediterraneo nell'età di Filippo II.* Turin: Einaudi.

———. 1986. *La Méditerranée: l'espace, l'histoire, les hommes et l'heritage.* Paris: Flammarion. **Italian translation**: 1992. *Il Mediterraneo: lo spazio, la storia, gli uomini, le tradizioni.* Milan: Bompiani.

Brent, Joseph. 1993. *Charles Sanders Peirce: A Life.* Bloomington: Indiana University Press.

Brusatin, Manlio. 1993. *Storia delle line.* Turin: Einaudi.

Buchanan, Colin. 1963. *Traffic in Towns.* Harmondsworth: Penguin Books.

Bull, William E., and Harry F. Williams (eds). 1959. *Semeiança del Mundo: A Medieval Description of the World.* Berkeley and Los Angeles: University of California Press.

Bunge, William. 1962. *Theoretical Geography.* Lund: Gleerup.

———. 1969. 'The First Years of the Detroit Geographical Expedition: A Personal Report'. *Field Notes* 1: 1–9.

Burgess, Ernest. 1925. 'Urban Areas' in Thomas Vernor Smith, Leonard Dupee White (eds), *Chicago: An Experiment in Social Science Research.* Chicago: University of Chicago Press, pp. 113–18.

BURTON, Ian. 1963. 'The Quantitative Revolution and Theoretical Geography'. *The Canadian Geographer* 2(7): 13–23.

BURTON RUSSELL, Jeffrey. 1991. *Inventing the Flat Earth: Columbus and Modern Historians*. New York: Praeger.

CAMILLE, Michael. 2000. 'Signs of the City: Place, Power, and Public Fantasy in Medieval Paris' in Barbara A. Hanawalt and Michal Kobialka (eds), *Medieval Practices of Space*. Minneapolis: University of Minnesota Press, pp. 1–36.

CAPOT-REY, Robert. 1946. *Géographie de la circulation sur les continents*. Paris: Gallimard.

CARETTI, Lanfranco. 1961. *Ariosto e Tasso*. Turin: Einaudi.

CASATI, Roberto. 2000. *La scoperta dell'ombra*. Milan: Mondadori.

CASSINI DE THURY, César-François. 1749. 'Sur la description géometrique de la France' in *Histoire de l'Académie Royale des Sciences*. Paris: Imprimerie Royale, pp. 553–60.

CASSIRER, Ernst. 1918. *Kants Leben und Lehre*. Hamburg: Felix Meiner Verlag. **Italian translation**: 1977. *Vita e dottrina di Kant*. Firenze: La Nuova Italia.

CASTELLS, Manuel. 1989. *The Informational City: Information Technology, Economic Restructuring, and the Urban–Regional Process*. Oxford: Blackwell.

––––. 1996. *The Rise of the Network Society*. Oxford: Blackwell. **Italian translation**: 2002. *La nascita della società in rete*. Milan: Università Bocconi.

––––. 1997. *The Power of Identity*. Oxford: Blackwell. **Italian translation**: 2003. *Il potere delle identità*. Milan: Università Bocconi.

CATTANEO, Carlo. 1972. 'La città considerata come principio ideale delle storie italiane' (1858) in *Opere scelte* (D. Castelnuovo Frigessi ed.), VOL. 4. Turin: Einaudi, pp. 79–126.

CAVALLI-SFORZA, Luigi Luca, Paolo Menozzi and Alberto Piazza. 1994. *The History and Geography of Human Genes*. Princeton, NJ: Princeton University Press. **Italian translation**: 1997. *Storia e Geografia dei Geni Umani*. Milan: Adelphi.

CHATEAUBRIAND, François-René de. 1850[1811]. 'Itinèraire de Paris à Jérusalem et de Jérusalem à Paris' in *Oeuvres complètes*, VOL. 5. Paris: A. Tétot, pp. 362–482.

CHEVALLIER, Raymond. 1980. *La romanisation de la Celtique du Pô, Volume 1: Les données géographiques*. Paris: Les Belles Lettres.

CHORLEY, Richard J., Robert P. Beckinsale and Antony J. Dunn. 1973. *The History of the Study of Landforms; Or, the Development of Geomorphology, Volume 2: The Life and Work of William Morris Davis*. London: Methuen.

CLAUSEWITZ, Carl Philipp Gottfried von. 1832. *Vom Kriege*. Berlin: F. Dummler. **Italian translation:** 1970. *Della guerra*. Milan: Mondadori.

CLAVAL, Paul. 1964. *Essai sur l'évolution de la géographie humaine*. Besançon: University of Besançon. **Italian translation:** 1972. *L'evoluzione storica della geografia umana*. Milan: Angeli.

CLIFFORD, James. 1997, *Routes: Travel and Translation in the Late Twentieth Century*. Cambrdige, MA: Harvard University Press. **Italian translation:** 1999. *Strade. Viaggio e traduzione alla fine del secolo XX*. Turin: Bollati Boringhieri.

COMBA, Rinaldo. 1993. 'I borghi nuovi dal progetto alla realizzazione' in Rinaldo Comba and Aldo Angelo Settia (eds), *I borghi nuovi: secoli xii-xiv*. Cuneo: Società per gli Studi Storici, Archeologici e Artistici della Provincia di Cuneo, pp. 279–300.

COREY, Kenneth E. 1982. 'Transactional Forces and the City'. *Ekistics* 49: 416–23.

COSGROVE, Denis. 1984. *Social Formation and Symbolic Landscape*. London: Croom Helm. **Italian translation:** 1990. *Realtà sociali e paesaggio simbolico*. Milan: Unicopli.

DAINELLI, Giotto. 1933. 'Le ragioni geografiche di una civiltà europea unitaria'. *Bollettino della Reale Società Geografica Italiana* 70: 3–28.

DAINVILLE, François de. 1962. 'De la profondeur à l'altitude. Des origines marines de l'expression cartographique du relief terrestre par côtes et courbes de niveau'. *Internationale Jahrbuch für Kartographie* 2: 151–62.

———. 1964, *Le langage des géographes: termes, signes, couleurs des cartes anciennes (1500–1800)*. Paris Picard et Cie.

DAMÁSIO, António R. 1994. *Descartes' Error: Emotion, Reason and the Human Brain*. New York: Putnam. **Italian translation:** 1995. *L'errore di Cartesio. Emozione, ragione e cervello umano*. Milan: Adelphi.

DANIEL, Hermann Adelbart (ed.). 1862. *Allgemeine Erdkunde. Vorlesungen an der Universität zu Berlin gehalten von Carl Ritter*. Berlin: Reimer.

DAVIS, Mike. 1990. *City of Quartz: Excavating the Future in Los Angeles*. London: Verso. **Italian translation:** 1999. *Città di quarzo: indagando sul futuro a Los Angeles*. Rome: Manifestolibri.

DEFFONTAINES, Pierre. 1938. *Le Brésil*. Paris: Colin.

———. 1972. *El Mediterraneo. La tierra, el mar, los hombres*. Barcelona: Editorial Juventud.

DELANO-SMITH, Catherine. 1987. 'Cartography in the Prehistoric Period in the Old World: Europe, the Middle East, and North Africa' in J. B. Harley and David Woodward (eds), *The History of Cartography, Volume 1: Cartography in Prehistoric, Ancient, and Medieval Europe and the Mediterranean*. Chicago: University of Chicago Press, pp. 54–101.

DELEUZE, Gilles. 1968. *Différence et repetition* (Difference and Repetition). Paris: Presses Universitaires de France. **Italian translation**: 1997. *Differenza e ripetizione*. Milan: Cortina.

———. 1988. *Le pli. Leibniz et le Baroque*. Paris: Éditions de Minuit. **Italian translation**: 1990. *La piega. Leibniz e il Barocco*. Turin: Einaudi.

DEMANGEON, Albert. 1942. *Problèmes de géographie humaine*. Paris: Colin.

DEMATTEIS, Giuseppe. 1970. *'Rivoluzione quantitativa' e nuova geografia*. Turin: Laboratorio di Geografia Economica dell'Università.

———. 1978. 'La crisi della città contemporanea' in Touring Club Italiano (ed.), *Le città*. Milan: Touring Club Italiano, pp. 179–91.

DERRUAU, Max. 1961. *Précis de géographie humaine*. Paris: Colin.

DE SETA, Cesare. 1976. 'Bari' in Lucio Gambi and Giulio Bollati (eds), *Storia d'Italia, Volume 6: Atlante*. Turin: Einaudi, pp. 405–07.

DESPLANQUES, Henri. 1959. 'Il paesaggio rurale della coltura promiscua in Italia'. *Rivista geografica italiana* 66: 29–64.

———. 1969. *Campagnes ombriennes. Contribution à l'étude des paysages ruraux en Italie Centrale*. Paris: Colin. **Italian translation**: 1975. *Campagne umbre, Regione dell'Umbria*. Perugia: Guerra.

DETIENNE, Marcel, and Jean-Pierre Vernant. 1974. *Les ruses de l'intelligence. La métis des Grecs*. Paris: Flammarion.

DEVLIN, Keith. 1997. *Goodbye Descartes: The End of Logic and the Search for a New Cosmology of the Mind*. New York: Wiley and Sons. **Italian translation**: 1999. *Addio, Cartesio. La fine della logica e la ricerca di una nuova cosmologia della mente*. Turin: Bollati Boringhieri.

DIAMOND, Cora (ed.). 1976. *Wittgenstein's Lectures on the Foundations of Mathematics: Cambridge, 1939*. Hassocks: Harvester Press.

DIAMOND, Jared. 1997. *Guns, Germs, and Steel: The Fates of Human Societies*. New York: Norton. **Italian translation**: 1998. *Armi, acciaio e malattie. Breve storia del mondo negli ultimi tredicimila anni*. Turin: Einaudi.

DIELS, Hermann, and Walther Kranz. 1922. *Die Fragmente der Vorsokratiker*. Berlin: Weidmann. **Italian translation**: 1986. *I Presocratici. Testimonianze e frammenti*. Bari: Laterza.

DUNCAN, James S. 1930. *The City as Text: The Politics of Landscape Interpretation in the Kanjan Kingdom*, Cambridge: Cambridge University Press.

DUYVENDAK, J. J. L. (ed.). 1953. *Tao Tö King. Le Livre de la Voie et de la Vertu*. Paris: Adrien-Maissoneuve. **Italian translation**: 1988. *Tao-Tê-Ching. Il Libro della Via e della Virtú*. Milan: Bompiani.

DUPONT, Alphonse. 1946. 'Espace et humanisme'. *Bibliothèque de Humanisme et Renaissance* 8: 220–93. **Italian translation**: 1993. *Spazio e Umanesimo. L'invenzione del Nuovo Mondo*. Venice: Marsilio.

DUPUY, Jean-Pierre. 1982. *Ordres et désordres. Enquête sur un nouveau paradigme*. Paris: Seuil.

ECO, Umberto. 1973. *Il Segno*. Milan: Isedi.

——. 1989. 'The Meaning of "The Meaning of the Meaning"', introduction to C. K. Ogden and I. A. Richards, *The Meaning of Meaning: A Study of the Influence of Language upon Thought and of the Science of Symbolism*. San Diego, CA: Harcourt Brace Jovanovich, pp. *v–xi*.

——. 1990. *I limiti dell'interpretazione*. Milan: Bompiani.

EDGERTON, Samuel Y. 1975. *The Renaissance Rediscovery of Linear Perspective*. New York: Basic Books.

EDSON, Evelyn. 1997. *Mapping Time and Space: How Medieval Mapmakers Viewed Their World*. London: British Library.

EICHENGREEN, Barry. 1996. *Globalizing Capital: A History of the International Monetary System*. Princeton, IL: Princeton University Press. **Italian translation**: 1998. *La globalizzazione del capitale. Storie del sistema monetario internazionale*. Milan: Baldini e Castoldi.

EMMER, Pieter C. 1993. '"Wir sind hier, weil ihr dort wart". Europäischer Kolonialismus und interkoloniale Migration'. *Concilium* 248: 304–12.

FARA, Amelio. 1993. *La città da Guerra*. Turin: Einaudi.

FARINELLI, Franco. 1976. 'La cartografia della campagna nel Novecento' in Lucio Gambi and Giulio Bollati (eds), *Storia d'Italia, Volume 6: Atlante*. Turin: Einaudi, pp. 626–54.

———. 1977. 'La casa rurale nel Medio Indostan' *Rivista Geografica Italiana* 84: 73–100.

———. 1980. 'Come Lucien Febvre inventò il possibilismo', introduction to Lucian Febvre, *La Terra e l'evoluzione umana. Introduzione geografica alla storia*. Turin: Einaudi, pp. *xi–xxxvii*.

———. 1981a. 'Crisi e critica della geografia borghese: il soggetto, l'oggetto, il terreno' in Franca Canigiani, Maria Carazzi and Eduardo Grottanelli (eds), *L'inchiesta sul terreno in geografia*, Turin: Giappichelli, pp. 49–58.

———. 1981b. 'Il villaggio indiano, o della geografia delle sedi: una critica' in Franco Farinelli (ed.), *Il villaggio indiano. Scienza, ideologia e geografia delle sedi*. Milan: Angeli, pp. 9–50.

———. 1983. 'Alle origini della geografia politica "borghese"' in Claude Raffestin (ed.), *Geografia politica. Teorie per un progetto sociale*. Milan: Unicopli, pp. 21–38.

———. 1984. *I lineamenti geografici della conurbazione lineare emilianoromagnola*. Bologna: Istituto di Geografia dell'Università.

———. 1985. 'La "Gerusalemme" catturata: ipotesi per una geografia del Tasso' in Andrea Buzzoni (ed.), *Torquato Tasso tra letteratura, musica, teatro e arti figurative*. Bologna: Nuova Alfa Editoriale, pp. 75–82.

———. 1986. 'Luoghi, strade, spazio: tra cartografia, geografia e potere'. *Urbanistica* 84 (August): 6–19.

———. 1989a. 'Certezza del rappresentare'. *Urbanistica* 97 (December): 7–16.

———. 1989b. *Pour une théorie générale de la géographie*, Geneva: Department of Geography, University of Geneva.

———. 1991. 'L'arguzia del paesaggio'. *Casabella* 575–6 (January–February): 10–12.

———. 1992. *I segni del mondo. Immagine cartografica e discorso geografico in età moderna*. Firenze: La Nuova Italia.

———. 1994. 'Squaring the Circle, or the Nature of Political Identity' in Franco Farinelli, Gunnar Olsson and Dagmar Reichert (eds), *Limits of Representation*. Munich: Accedo, pp. 11–28.

——. 1995. 'Per una nuova geografia del Mediterraneo' in Lorenzo Bellicini (ed.), *Mediterraneo. Città, territorio, economie alle soglie del xxi secolo*. Rome: Cresme, pp. 121–48.

——. 1997. 'L'immagine dell'Italia' in Pasquale Coppola (ed.), *Geografia politica delle regioni italiane*. Turin: Einaudi, pp. 33–59.

——. 1998. 'Did Anaximander ever Say (or Write) any Words? The Nature of Cartographical Reason'. *Ethics, Place and Environment* 1(2): 135–44.

——. 2000. 'I caratteri originali del paesaggio abruzzese' in Massimo Costantini and Costantino Felice (eds), *Storia d'Italia. Le regioni dall'Unità ad oggi*. Turin: Einaudi, pp. 121–54.

——. 2002. 'Il mondo, la mappa, il labirinto' in Gianluca Bocchi and Mauro Ceruti (eds), *Origini della scrittura. Genealogie di un'invenzione*. Milan: Bruno Mondadori, pp. 225–34.

FAUCONNIER, Gilles. 1997. *Mappings in Thought and Language*. Cambridge: Cambridge University Press.

FAWCETT, C. B. 1932. 'Distribution of the Urban Population in Great Britain, 1931'. *Geographical Journal* 79(2): 100–13.

FEBVRE, Lucien. 1922, *La terre et l'évolution humaine. Introduction géographique à l'histoire*. Paris: Albin Michel. **Italian translation**: 1980. *La terra e l'evoluzione umana. Introduzione geografica alla storia*. Turin: Einaudi.

FIORINI, Matteo. 1881. *Le proiezioni delle carte geografiche*. Bologna: Zanichelli.

FISCHER, Wolfram. 1987. 'Wirtschaft, Gesellschaft und Staat in Europa, 1914–1980' in Wolfram Fischer (ed.), *Handbuch der europäischen Wirtschaft-und Sozialgeschichte*, VOL. 6. Stuttgart: Steiner, pp. 10–221.

FLORENSKIJ, Pavel. 1967. 'Obratnaja perspektiva'. *Trudy po znakovym sistemam* 3(198): 381–416. **Italian translation**: 1983. 'La prospettiva rovesciata' in Nicoletta Misler (ed.), *La prospettiva rovesciata e altri scritti*. Rome: Casa del Libro, pp. 73–135.

FOCILLON, Henri. 1943. *Vie des Formes*. Paris: Presses Universitaires de France. **Italian translation**: 1987. *Vita delle forme*. Turin: Einaudi.

FORREST, William George. 1963. *The Emergence of Greek Democracy: The Character of Greek Politics, 800–400 BC*. Oxford: Oxford University Press. **Italian translation**: 1966. *Le origini della democrazia greca. Caratteri del pensiero politico greco tra l'800 e il 400 a.C.* Milan: Il Saggiatore.

FRANCHEVILLE, M. de. 1752. *Le Siècle de Louis XIV*, 2 VOLS. Dresden: G. Conrad Walther.

FREGE, Gottlob. 1892. 'Uber Sinn Und Bedeutung' in *Zeitschrift für Philosophie und Philosophiche Kritik*, pp. 25–50. **Italian translation**: 1970. 'Senso e denotazione' in C. Lazzerini (ed.), *Ricerche logiche*. Bologna: Calderini, pp. 135–60.

FRIEDMAN, David. 1988. *Florentine New Towns: Urban Design in the Late Middle Ages*. Cambridge, MA: The MIT Press. **Italian translation**: 1996. *Terre nuove. La creazione delle città fiorentine nel tardo medioevo*. Turin: Einaudi.

FRIEDMANN, John. 1973. 'The Urban Field as Human Habitat' in S. P. Snow (ed.), *The Place of Planning*. Auburn, AL: Auburn University Printing Service, pp. 42–52.

———, and John Miller. 1965. 'The Urban Field'. *Journal of the American Institute of Planners* 31(4): 312–20.

GALILEI, Galileo. 1965[1623]. *Il Saggiatore*. Milan: Feltrinelli.

GALLOIS, Lucien. 1909. 'L'Académie des Sciences et les origines de la carte de Cassini'. *Annales de Géographie* 18: 193–204, 289–307.

GAMBI, Lucio. 1972. 'I valori storici dei quadri ambientali' in Ruggiero Romano and E. Corrado Vivanti (eds), *Storia d'Italia, Volume 1: I caratteri originali*. Turin: Einaudi, pp. 3–60.

———. 1973a. 'Critica ai concetti geografici di paesaggio umano' in *Una geografia per la storia*. Turin: Einaudi, pp. 148–74.

———. 1973b. 'Da città ad area metropolitana' in Ruggiero Romano and Corrado Vivanti (eds), *Storia d'Italia. Annali, Volume 5: I documenti*. Turin: Einaudi, pp. 365–424.

GARGANI, Aldo G. 1971. *Hobbes e la scienza*, Turin: Einaudi.

GARREAU, Joel. 1991. *Edge City: Life in the New Frontier*. New York: Doubleday.

GAUCHET, Marcel. 1985. *Le désenchantement du monde. Une histoire politique de la religion*. Paris: Gallimard.

GEDDES, Patrick. 1915. *Cities in Evolution: An Introduction to the Town Planning Movement and the Study of Civics*. London: Williams and Norgate. **Italian translation**: 1970. *Città in evoluzione*. Milan: Il Saggiatore.

GEORGE, Pierre. 1959. *Questions de Géographie de la Population*. Paris: Presses Universitaires de France. **Italian translation**: 1962. *Manuale di geografia della popolazione*. Milan: Edizioni di Comunità.

——. 1966. *Sociologie et géographie*. Paris: Presses Universitaires de France. **Italian translation**: 1976. *Geografia e sociologia*. Milan: Il Saggiatore.

——. 1970. *Dictionnaire de la Géographie*. Paris: Presses Universitaires de France. **Italian translation**: 1974. *Dizionario di Geografia*. Rome: Cesviet.

——. 1978. 'La città media nella megalopoli' in C. Muscarà (ed.), *Megalopoli mediterranea*. Milan: Angeli, pp. 134–45.

GERBI, Antonello. 1955. *La disputa del Nuovo Mondo. Storia di una polemica 1750–1900*. Milan: Ricciardi.

GILBERT, Walter. 1992. 'A Vision of the Grail' in Daniel J. Kevles and Leroy Hood (eds), *The Code of Codes: Social Issues in the Human Genome Project*. Cambridge, MA: Harvard University Press, pp. 83–97.

GILLIES, James, and Robert Cailliau. 2000. *How the Web Was Born*. Oxford: Oxford University Press. **Italian translation**: 2002. *Com'è nato il Web*. Milan: Baldini and Castoldi.

GLACKEN, Clarence J. 1967. *Traces on the Rhodian Shore*. Berkeley: University of California Press.

GOETHE, Johann Wolfgang von. 1790. 'Torquato Tasso' in *Goethes Schriften*, VOL. 6. Leipzig: Göschen, pp. 1–222. **Italian translation**: 1988. *Torquato Tasso*. Venice: Marsilio.

——. 1903. 'Italianische Reise' in *Goethes Werke*, VOL. 30. Weimar: Böhlau. **Italian translation**: 1975. *Ricordi di Viaggio in Italia, 1786–88*, VOL. 1. Firenze: Sansoni.

GOLD, John R. 1980. *An Introduction to Behavioural Geography*. Oxford: Oxford University Press. **Italian translation**: 1985. *Introduzione alla geografia del comportamento*. Milan: Angeli.

GOLDFINGER, Charles. 1986. *La géofinance: Pour comprendre la mutation financière*. Paris: Seuil.

GOTTMANN, Jean. 1957. 'Megalopolis, or the Urbanization of the Northeastern Seaboard of the United States'. *Economic Geography* 33(3): 189–98.

——. 1961. *Megalopolis: The Urbanized Northeastern Seaboard of the United States*. New York: The Twentieth Century Fund. **Italian translation**: 1970. *Megalopoli. Funzioni e relazioni di una pluri-città*, Turin: Einaudi.

——. 1976a. 'Office Work and the Evolution of Cities'. *Ekistics* 46: 1–5.

——. 1976b. 'The Recent Evolution of Oxford'. *Ekistics* 46: 31–6.

——. 1978. 'Verso una megalopoli della pianura padana' in C. Muscarà (ed.), *Megalopoli mediterranea*. Milan: Angeli, pp. 23–31.

——. 1982. 'The Metamorphosis of the Modern Metropolis'. *Ekistics* 49: 7–11.

GOUROU, Pierre. 1973. *Pour une géographie humaine*. Paris: Flammarion.

——. 1982. *Terres de bonne espérance: Le monde tropical*. Paris: Plon.

GRAMSCI, Antonio. 1975[1948–51]. *Quaderni del carcere*, 3 VOLS. Turin: Einaudi.

GRESH, Alain, and Philippe Rekacewicz. 2000. 'Le carte dei negoziati arabo israeliani'. *Le Monde Diplomatique* (Italian edition) 7(2): 12–13.

GUIDONI, Enrico. 1970. *Arte e urbanistica in Toscana, 1000–1315*. Rome: Bulzoni.

HARBISON, Robert. 2000. *Reflections on Baroque*. London: Reaktion Books.

HARD, Gerhard. 1969. '"Dunstige Klarheit". Zu Goethes Beschreibung der italienischen Landschaft'. *Die Erde* 2–4: 138–54.

HARRIS, Chauncey D., and Edward L. Ullmann. 1945. 'The Nature of Cities'. *The Annals of the American Academy of Political and Social Science* 242: 7–17.

HARTSHORNE, Richard. 1939. *The Nature of Geography. A Critical Survey of Current Thought in the Light of the Past*. Lancaster, PA: The Association of American Geographers.

HARTSHORNE, Charles, and Paul Weiss (eds). 1978. *Collected Papers of Charles Sanders Peirce, Volume 2: Elements of Logic*. Cambridge, MA: Belknap Press.

HARVEY, David. 1985. *The Urbanization of Capital*. Oxford: Blackwell.

HARVEY, P. D. A. 1985. 'The Spread of Mapping to Scale in Europe, 1500–1550' in Carla Clivio Marzoli, Giacomo Corna Pellegrini and Gaetano Ferro (eds), *Imago et Mensura Mundi, Atti del IX Congresso Internazionale di Cartografia*. Rome: Istituto dell'Enciclopedia Italiana, pp. 473–7.

HASSINGER, Hugo. 1931. *Geographische Grundlagen der Geschichte*. Freiburg: Herder.

HEGEL, G. W. F. 1955. *Aesthetik*. Berlin: Aufbau. **Italian translation**: 1963. *Estetica*. Milan: Feltrinelli.

——. 1996[1837]. *Vorlesungen über die Philosophie der Geschichte*. Hamburg: Meiner. **Italian translation**: 2001. *Filosofia della storia universal*. Turin: Einaudi.

HEIDEGGER, Martin. 1950. *Holzwege*. Frankfurt am Main: Klostermann. **Italian translation**: 1968. *Sentieri interrotti*. Firenze: La Nuova Italia.

———. 1954. *Vorträge und Aufsätze*. Pfullingen: Neske. **Italian translation**: 1976. *Saggi e discorsi*. Milan: Mursia.

———. 1981. *Erläuterungen zu Hölderlins Dichtung*. Frankfurt am Main: Klostermann. **Italian translation**: 1988. *La poesia di Hölderlin*. Milan: Adelphi.

HETTNER, Alfred. 1923. 'Methodische Zeit und Streitfragen'. *Geographische Zeitschrift* 39: 37–60.

HOBBES, Thomas. 1951[1651]. *Leviathan*. Harmondsworth: Penguin. **Italian translation**: 1989. *Leviatano*. Bari: Laterza.

HOFSTADTER, Douglas R. 1979. *Gödel, Escher, Bach: An Eternal Golden Braid*. New York: Basic Books. **Italian translation**: 1984. *Gödel, Escher, Bach: un'eterna ghirlanda brillante*. Milan: Adelphi.

HOLDICH, Thomas. 1902. 'Some Geographical Problems'. *The Geographical Journal* 20: 411–27.

HORKHEIMER, Max, and Theodor W. Adorno. 1947. *Dialektik der Aufklärung. Philosophische Fragmente*. Amsterdam: Querido. **Italian translation**: 1966. *Dialettica dell'illuminismo*. Turin: Einaudi.

HOUËL, Jean-Pierre. 1782. *Voyage pittoresque des Isles de Sicile, de Malte et de Lipari, ou l'on traite des antiquités qui s'y trouvent ancore; des principaux phènomènes que la nature y offre; du costume des habitans, et de quelques usages*. Paris: de l'Imprimerie de Monsieur.

HOYT, Homer. 1939. *The Stucture and Growth of Residential Neighbourhoods in American Cities*. Washington, DC: Federal Housing Administration, USA.

HÜBNER, Kurt. 1985. *Die Wahrheit des Mythos. Mythische Welterfahrungen im wissenshaftlichen Zeitalter*. Munich: Beck'sche Verlagsbuchhandlung. **Italian translation**: 1990. *La verità del mito*. Milan: Feltrinelli.

HUGILL, Peter J. 1993. *World Trade since 1431: Geography, Technology, and Capitalism*. Baltimore, MD: Johns Hopkins University Press.

———. 1999. *Global Communications since 1844: Geopolitics and Technology*. Baltimore, MD: Johns Hopkins University Press.

HUMBOLDT, Alexander von. 1845. *Kosmos. Entwurf einer physischen Weltbeschreibung*, VOL. 1. Stuttgart: Cotta.

———. 1849. 'Die Lebenskraft oder der Rhodische Genius' in *Ansichten der Natur*. Berlin: Cotta, pp. 319–25. **Italian translation**: 1988, 'La forza vitale o il genio di Rodi' in *Quadri della Natura*. Firenze: La Nuova Italia, pp. 311–18.

Husserl, Edmund. 1954. *Die Krisis der europäischen Wissenschaften und die transzendentale Phänomenologie.* The Hague: Nijhoff. **Italian translation:** 1961. *La crisi delle scienze europee e la fenomenologia trascendentale.* Milan: Il Saggiatore.

Ivins, William M. 1985. 'La rationalisation du regard'. *Culture Technique* 14 (June): 31–7.

Jacobs, Jane. 1961. *The Death and the Life of Great American Cities.* New York: Random House. **Italian translation:** *Vita e morte delle grandi città. Saggio sulle metropoli americane.* Turin: Edizioni di Comunità.

———. 1969. *The Economy of Cities.* New York: Random House.

João, Elsa Maria. 1998. *Causes and Consequences of Map Generalisation.* London: Taylor and Francis.

Kant, Immanuel. 1932. 'Physische Geographie' in *Gesammelte Schriften*, VOL. 9. Berlin: Der Königlich Reussischen Akademie der Wissenschaften, pp. 156–65. **Italian translation:** 1807. *Geografia fisica*, VOL. 1. Milan: Giovanni Silvestri, pp. *xxii-iv.*

Kaplan, Robert. 1999, *The Nothing That Is: A Natural History of Zero.* Oxford: Oxford University Press. **Italian translation:** 1999. *Zero. Storia di una cifra.* Milan: Rizzoli.

Keller, Evelyn Fox. 2000. *The Century of the Gene.* Cambridge, MA: Harvard University Press. **Italian translation:** 2001. *Il secolo del gene.* Milan: Garzanti.

Kerényi, Karl. 1976, *Dyonisos. Urbild des unzerstörbaren Lebens.* Munich: Langen-Müller. **Italian translation:** 1992. *Dioniso. Archetipo della vita indistruttibile.* Milan: Adelphi.

Kern, Stephen. 1983. *The Culture of Time and Space, 1880–1918.* Cambridge, MA: Harvard University Press. **Italian translation:** 1988. *Il tempo e lo spazio. La percezione del mondo tra Otto e Novecento.* Bologna: Il Mulino.

King, Anthony D. 1990. *Global Cities: Post-Imperialism and the Internationalization of London.* London: Routledge.

Kleist, Heinrich von. 1810. 'Über das Marionettentheater' in *Werke*, VOL. 5. Berlin: Bong, pp. 74–83.

Koerner, Joseph Leo. 1990. *Caspar David Friedrich and the Subject of Landscape.* London: Reaktion Books.

KOLAKOWSKI, Leszek. 1966. *Filozofia pozytywistyczna* (*Od Hume'a do Kola Wieden-skiego*). Warsaw: Panstwowe Wydawnictwo Naukowe. **Italian translation:** 1974. *La filosofia del positivism.* Bari: Laterza.

KORMOSS, I. B. F. 1978. 'Qualche considerazione sugli aspetti statistici e dinamici e sulle prospettive delle formazioni megalopolitane' in C. Muscarà (ed.), *Megalopoli mediterranea.* Milan: Angeli, pp. 32–57.

KOTKIN, Joel. 2000. *The New Geography: How the Digital Revolution Is Reshaping the American Landscape.* New York: Random House.

KRAMER, Gustav. 1875. *Carl Ritter. Ein Lebensbild nach seinem handschriftlichen Nachlass.* Halle: Buchhandlung des Waisenhauses.

KRANZBERG, Melvin. 1985. 'The Information Age: Evolution or Revolution?' in Bruce R. Guile (ed.), *Information Technologies and Social Transformation.* Washington, DC: National Academy of Engineering, pp. 45–59.

LACOSTE, Yves. 1975. *La géographie, ça sert, d'abord, à faire la guerre.* Paris: Maspero. **Italian translation:** 1977. *Crisi della geografia. Geografia della crisi.* Milan: Angeli.

———. 1982. 'Les deux Méditerranées'. *Hérodote* 7(27): 3–11.

LAMPARD, E. E. 1955. 'The History of Cities in the Economically Advanced Areas'. *Economic Development and Cultural Change* 3(2): 81–102.

LARNER, John. 1999. *Marco Polo and the Discovery of the World.* New Haven, CT: Yale University Press.

LEFEBVRE, Henri. 1974. *La production de l'espace.* Paris: Anthropos. **Italian translation:** 1976. *La produzione dello spazio.* Milan: Mozzi.

LEIBNIZ, Gottfried Wilhelm. 1961. 'Discours préliminaire de la conformità de la foy avec la raison' in Karl Immanuel Gerhardt (ed.), *Die Philosophischen Schriften von Gottfried Wilhelm Leibniz*, VOL. 2. Hildesheim: Olms, pp. 49–101. **Italian translation:** 1967. 'Discorso preliminare sulla conformità della fede con la ragione' in *Scritti filosofici*, VOL. 1. Turin: Utet, pp. 401–56.

LE LANNOU, Maurice. 1949. *La géographie humaine.* Paris: Flammarion.

LE ROY, Jean. 1935. '"Gross Berlin" Le Grand Berlin'. *Annales de Géographie* 44: 633–40.

LÉVÊQUE, Pierre, and Pierre Vidal-Naquet. 1964. *Clisthène l'Athénien. Essai sur la réprésentation de l'espace et du temps dans la pensée politique grecque de la fin du vie siècle à la mort de Platon.* Paris: Macula.

LÉVY, Pierre. 1995. *Qu'est ce que le virtuel?* Paris: La Découverte. **Italian translation:** 1997. *Il virtuale.* Milan: Cortina.

LEWIS, Martin W., and Karen Wigen. 1997. *The Myth of Continents: A Critique of Metageography.* Berkeley: University of California Press.

LEWONTIN, Richard C. 1992. 'The Dream of the Human Genome'. *The New York Review of Books* 39(9) (28 May): 31–40.

LIM, Lin Lean, and Nana Oishi. 1996. *International Labour Migration of Asian Women: Distinctive Characteristics and Policy Concerns.* Geneva: International Labour Office.

LIVET, Pierre. 1983. 'La fascination de l'auto-organisation' in Paul Dumouchel and Jean-Pierre Dupuy (eds), *L'auto-organisation. De la Physique au Politique.* Paris: Seuil, pp. 165–71.

LOPEZ, Robert S. 1963. 'The Crossroads within the Walls' in Oscar Handlin and John E. Burchard (eds), *The Historian and the City.* Cambridge, MA: The MIT Press, pp. 5–17.

LÜDDE, Johann Gottfried. 1849. *Die Geschichte der Methodologie der Erdkunde.* Leipzig: Hinrich.

MACKINDER, Halford John. 1904. 'The Geographical Pivot of History'. *The Geographical Journal* 23(4): 421–37. **Italian translation:** 1996. 'Il perno geografico della storia'. *I castelli di Yale* 1: 129–50.

———. 1943. 'The Round World and the Winning of the Peace'. *Foreign Affairs* 21: 595–605. **Italian translation:** 1994. 'Il mondo intero e come vincere la pace'. *Limes* 1: 171–82.

MALDONADO, Tomás. 1971. *La speranza progettuale. Ambiente e società.* Turin: Einaudi.

MALÉCOT, Gustave. 1969. *The Mathematics of Heredity* (Demetrios M. Yermanos trans.). San Francisco: W. H. Freeman.

MANDELBROT, Benoît. 1987. *La geometria della natura.* Milan: Montedison Progetto Cultura.

MARAZZI, Christian. 1994. *Il posto dei calzini. La svolta linguistica dell'economia e i suoi effetti sulla politica.* Bellinzona: Casagrande.

MARINELLI, Olinto. 1902. 'Alcune questioni relative al moderno indirizzo della geografia'. *Rivista Geografica Italiana* 9: 217–40.

MASSEY, Doreen, and Pat Jess (eds). *A Place in the World? Places, Cultures and Globalization.* Oxford: Oxford University Press. **Italian translation**: 2001. *Luoghi, culture e globalizzazione.* Turin: Utet.

MATTELART, Armand. 1994. *L'invention de la communication.* Paris: La Découverte. **Italian translation**: 1998. *L'invenzione della comunicazione. La via delle idee.* Milan: Il Saggiatore.

MATURANA, Humberto, and Francisco Varela. 1980. *Autopoiesis and Cognition: The Realisation of Living.* Dordrecht: Reidel. **Italian translation**: 1985. *Autopoiesi e cognizione. La realizzazione del vivente.* Venice: Marsilio.

MCKENZIE, Roderick Duncan. 1933. *The Metropolitan Community.* New York: McGraw Hill.

MCLUHAN, Marshall. 1962. *The Gutenberg Galaxy: The Making of Typographic Man.* Toronto: Toronto University Press. **Italian translation**: 1976. *La Galassia Gutenberg. Nascita dell'uomo tipografico.* Rome: Armando.

——, and Bruce R. Powers. 1989. *The Global Village: Transformations in World Life and Media in the 21st Century.* Oxford: Oxford University Press. **Italian translation**: 1992. *Il Villaggio Globale. xxi secolo: trasformazioni nella vita e nei media.* Milan: SugarCo.

MEHRING, Franz. 1947[1910]. *Deutsche Geschichte vom Ausgang des Mittelalters.* Berlin: Dietz. **Italian translation**: 1957. *Storia della Germania moderna.* Milan: Feltrinelli.

MEIER, Christian. 1980. *Die Entstehung des Politischen bei den Griechen.* Frankfurt am Main: Suhrkamp. **Italian translation**: 1988. *La nascita della categoria del politico in Grecia.* Bologna: Il Mulino.

MEIER, Richard L. 1962. *A Communication Theory of Urban Growth.* Cambridge, MA: The MIT Press. **Italian translation**: 1962. *Teoria della comunicazione e struttura urbana.* Milan: Il Saggiatore.

MELLAART, James. 1967. *Çatal Hüyük: A Neolithic Town in Anatolia.* London: Thames and Hudson.

MENOZZI, Paolo, Alberto Piazza and Luigi Luca Cavalli-Sforza. 1978. 'Synthetic Maps of Human Gene Frequencies in Europe'. *Science* 201(4358): 786–92.

MITCHELL, William J. 1995. *City of Bits: Space, Place, and the Infobahn.* Cambridge, MA: The MIT Press.

MITTERMEIER, Russell, Cristina Goettsch Mittermeier, Patricio Robles Gil, Gustavo Fonseca, Thomas Brooks, John Pilgrim and William R. Konstant. 2003. *Wilderness: Earth's Last Wild Places*. Arlington, VA: Conservation International.

MORACHIELLO, Paolo. 2003. *La città greca*. Rome: Laterza.

MORICE, A. 1997. 'I lavoratori stranieri agli avamposti della precarietà'. *Le Monde Diplomatique* (Italian edition) 4(1): 14–15.

MUMFORD, Lewis. 1938. *The Culture of Cities*. San Diego, CA: Harcourt, Brace. **Italian translation**: 1999. *La cultura delle città*. Turin: Edizioni di Comunità.

——. 1961. *The City in History*. New York: Harcourt Brace Jovanovich. **Italian translation**: 1963. *La città nella storia*. Milan: Comunità.

MYRES, John L. 1896. 'An Attempt to Reconstruct the Maps Used by Herodotus'. *The Geographical Journal* 8: 605–29. **Italian translation**: 1983. 'Erodoto geografo' in Francesco Prontera (ed.), *Geografia e geografi nel mondo antico*. Rome: Laterza, pp. 115–34.

NATOLI, Salvatore. 1996. *Soggetto e fondamento. Il sapere dell'origine e la scientificità della filosofia*. Milan: Bruno Mondadori.

NENCI, Giuseppe. 1979. 'Formazione e carattere dell'impero ateniese' in Ranuccio Bianchi Bandinelli (ed.), *Storia e civiltà dei Greci, Volume 3: La Grecia nell'età di Pericle. Storia, letteratura, filosofia*. Milan: Bompiani, pp. 45–92.

NEUMANN, Joachim. 1977. 'Über Begriffe der kartographischen Generalisierung'. *International Yearbook of Cartography* 17: 119–24.

NISSEN, Heinrich. 1902. *Italische Landeskunde*, VOL. 2, PART 1. Berlin: Weidmannsche Buchhandlung.

NUGENT, Walter T. K. 1992. *Crossings: The Great Transatlantic Migrations, 1870–1914*. Bloomington: Indiana University Press.

OGDEN, C. K., and I. A. Richards. 1989. *The Meaning of Meaning: A Study of the Influence of Language upon Thought and of the Science of Symbolism*. San Diego: Harcourt Brace Jovanovich.

OLSSON, Gunnar. 1974. 'The Dialectics of Spatial Analysis'. *Antipode* 6(3): 50–62. **Italian translation**: 1991. *Linee senza ombre. La tragedia della pianificazione*. Rome: Theoria.

——. 1980. *Birds in Egg / Eggs in Bird*. London: Pion. **Italian translation**: 1987. *Uccelli nell'uovo / Uova nell'uccello*. Rome: Teoria.

ORTOLANI, Mario. 1984. *Geografia delle sedi*. Padova: Piccin.

———. 1992. *Geografia della popolazione*. Padova: Piccin.

PANOFSKY, Erwin. 1927. 'Die Perspektive als "symbolische Form"' in Fritz Saxl (ed.), *Vorträge der Bibliothek Warburg*. Leipizig: Teubner, pp. 258–330. **Italian translation**: 1961. 'La prospettiva come "forma simbolica"' in *La prospettiva come 'forma simbolica' e altri scritti*. Milan: Feltrinelli, pp. 35–114.

PECORA, Aldo. 1970. 'La "corte" padana" in Giuseppe Barbieri and Lucio Gambi (eds), *La casa rurale in Italia*. Firenze: Olschki, pp. 219–44.

PERRIN, Jean. 1948. *Les atomes*. Paris: Presses Universitaires de France. **Italian translation**: 1981. *Gli atomi*. Rome: Editori Riuniti.

PESCHEL, Oscar. 1876. *Neue Probleme der vergleichenden Erdkunde als Versuche einer Morphologie der Erdoberflache*. Leipzig Duncker und Humblot,.

PINCHEMEL, Philippe. 1972. 'Paul Vidal de la Blache' in E. Meynen (ed.), *Geographische Taschenbuch, 1970–72*. Wiesbaden: Steiner, pp. 47–51.

PIZZORNO, Alessandro. 1967. 'Introduzione' in Robert Park, Ernest W. Burgess and Roderick Duncan McKenzie, *La città*. Milan: Comunità, pp. *v–xxiii*.

POULS, H. C. 1980. 'Mieux vaut voir que courir 2' in Giulio Macchi (ed.), *Cartes et figures de la Terre*. Paris: Centre Georges Pompidou, pp. 248–51.

PRED, Allan. 1977. *City-Systems in Advanced Economies: Past Growth, Present Processes and Future Development Options*. London: Hutchinson.

PRONTERA, Francesco. 1997. 'Sulle basi empiriche della cartografia greca'. *Sileno* 23: 49–64.

PUMAIN, Denise, Thérèse Saint-Julien and Léna Sanders. 1989. *Villes et auto-organisation*. Paris: Economica.

RACINE, Jean-Bernard. 1971. 'Le modèle urbain américain. Les mots et les choses'. *Annales de Géographie* 80: 397–427.

———. 1993. *La ville entre Dieu et les homes*. Paris: Anthropos.

RAFIULLAH, S. M. 1966. *The Geography of Transhumance*. Aligarh: Department of Geography, Aligarh Muslim University.

RAIMONDI, Ezio. 1980. 'Il dramma nel racconto. Topologia di un poema' in *Poesia come retorica*. Firenze: Olschki, pp. 71–202.

RAMBAUD, Michel. 1974, *L'espace dans le récit césarien*, in Raymond Chevallier (ed.), *Mélanges offerts à Roger Dion*. Paris: Picard, pp. 111–29.

RATZEL, Friedrich. 1899. *Anthropogeographie. Grundzüge der Anwendung der Geographie auf die Geschichte*. Stuttgart: Engelhorn. **Italian translation**: 1914[1911]. *Geografia dell'uomo. Principi d'applicazione della scienza geografica alla storia*. Turin: Fratelli Bocca.

REICHENBACH, Hans. 1957. *The Philosophy of Space and Time*. New York: Dover.

RITTER, Carl. 1852. *Einleitung zur allgemeinen vergleichenden Geographie, und Abhandlungen zur Begründung einer mehr wissenschaftlichen Behandlung der Erdkunde*. Berlin: Reimer.

———. 1974. *Introduction à la géographie générale comparée*. Paris: Les Belles Lettres.

RYKWERT, Joseph. 1988. *The Idea of a Town: The Anthropology of Urban Form in Rome, Italy and the Ancient World*. Cambridge, MA: The MIT Press. **Italian translation**: 2002. *L'idea di città. Antropologia della forma urbana nel mondo antico*. Milan: Adelphi.

ROMAGNOLI, Siamo. 1966, *Tasso*. Milan: Cei.

ROSEN, Joe. 1995. *Symmetry in Science: An Introduction to the General Theory*. New York: Springer.

ROSSI, Pietro. 1975. *Storia universale e geografia in Hegel*. Firenze: Sansoni.

SACK, Robert David. 1986. *Human Territoriality: Its Theory and History*. Cambridge: Cambridge University Press.

SANSONI, U. 1982. 'Il ciclo evolutivo della civiltà: relazioni fra arte rupestre e ambiente in Valcamonica' in Emmanuel Anati (ed.), *Il caso Valcamonica: rapporto uomo-territorio nella dinamica della storia*. Milan: Unicopli, pp. 59–71.

SANTILLANA, Giorgio de, and Hertha von Dechend. 1969. *Hamlet's Mill: An Essay Investigating the Origins of Human Knowledge and Its Transmission Through Myth*. Boston, MA: Gambit. **Italian translation**: 1983. *Il mulino di Amleto. Saggio sulmito e sulla struttura del tempo*. Milan: Adelphi.

SANTEL, Bernhard. 1995. *Migration in und nach Europa. Erfahrungen, Strukturen, Politik*. Opladen: Leske und Budrich.

SASSEN, Saskia. 1994. *Cities in a World Economy*. Thousand Oaks, CA: Pine Forge Press. **Italian translation**: 1997. *Le città nell'economia globale*. Bologna: Il Mulino.

SAUSSURE, Ferdinand de. 1916. *Cours de linguistique générale*. Lausanne: Payot. **Italian translation**: 1972. *Corsodi linguistica generale*. Bari: Laterza.

SCHIVELBUSCH, Wolfgang. 1977. *Geschichte der Eisenbahnreise*. Munich: Hanser. **Italian translation:** 1988. *Storia dei viaggi in ferrovia*, Turin: Einaudi.

——. 1899. 'Bemerkungen zur Siedelungsgeographie'. *Geographische Zeitschrift* 5: 59–71.

——. 1906. *Die Ziele der Geographie des Menschen*. Munich: Oldenbourg.

——. 1919. 'Die Stellung der Geographie des Menschen in der erdkundlichen Wissenschaft'. *Geographische Abende im Zentraleninstitut für Erziehung und Unterricht* 5: 15–20.

SCHMITT, Carl. 1938. *Der Leviathan in der Staatslehre des Thomas Hobbes. Sinn und Fehlschlag eines politischen Symbols*. Hamburg: Hanseatischen Verlagsanstalt.

——. 1974. *Der Nomos der Erde im Völkerrecht des Jus Publicum Europaeum*. Berlin: Duncker u. Humblot. **Italian translation:** 1991. *Il Nomos della Terra nel diritto internazionale dello 'Jus Publicum Europaeum'*. Milan: Adelphi.

SCHRÖDINGER, Erwin. 1944. *What Is Life*? Cambridge: Cambridge University Press. **Italian translation:** 1995. *Che cos'è la vita*? Milan: Adelphi.

SELVINI, Attolio, and Franco Guzzetti. 1999. *Cartografia generale: tematica e numerica*. Turin: Utet.

SERRES, Michel. 1993. *Les origines de la géometrie*. Paris: Flammarion. **Italian translation:** 1994. *Le origini della geometria*. Milan: Feltrinelli.

SHANNON, Claude E., and Warren Weaver. 1949. *The Mathematical Theory of Communication*. Glencoe: University of Illinois Press. **Italian translation:** 1971. *La teoria matematica delle comunicazioni*. Milan: Etas.

SLICHER VAN BATH, Bernard Hendrik. 1960. *De agrarische geschiedenis van West-Europa, 500-1850*. Utrecht: Het Spectrum. **Italian translation:** 1972. *Storia agraria dell'Europa occidentale, 500–1850*. Turin: Einaudi.

SMITH, Adam. 1776. *An Inquiry into the Nature and Causes of the Wealth of Nations*. London: Strahan and Cadell. **Italian translation:** 1948. *Ricerche sopra la natura e le cause della ricchezza delle nazioni*. Turin: Utet.

SMITH, Christine. 1994. 'L'occhio alato: Leon Battista Alberti e la rappresentazione di passato, presente e futuro' in Millon Henry and Magnago Lampugnani Vittorio (eds), *Rinascimento. Da Brunelleschi a Michelangelo: la rappresentazione dell'architettura*. Milan: Bompiani, pp. 453–55.

SNYDER, John P. 1993. *Flattening the Earth: Two Thousand Years of Map Projections*. Chicago: University of Chicago Press.

SOJA, Edward W. 2000. *Postmetropolis: Critical Studies of Cities and Regions*. Oxford: Blackwell.

SORRE, Max. 1948. 'La notion de genre de vie et sa valeur actuelle'. *Annales de Géographie* 57: 97–108, 193–204.

———. 1952. *Les fondements de la géographie humaine, Volume 3: L'habitat; Conclusion générale*. Paris: Colin.

———. 1955. *Les migrations des peuples. Essai sur la mobilité géographique*. Paris: Flammarion.

———. 1957. *Rencontres de la géographie et de la sociologie*. Paris: Rivière.

STAROBINSKI, Jean. 1992. 'L'imitation du Tasse'. *Annales de la Société Jean-Jacques Rousseau* 40: 265–88. **Italian translation**: 1994. *Rousseau e Tasso*, Turin: Bollati Boringhieri.

STAVENHAGEN, W. 1900. 'Die geschichtliche Entwickelung des preussischen Militärkarten Wesens'. *Geographische Zeitschrift* 6: 504–12.

STORK, J. 1996. 'Bahrain in lotta per la democrazia'. *Le Monde Diplomatique* (Italian edition) 3(7): 13.

STRAVINSKY, Igor. 1935. *Chroniques de ma vie*. Paris: Denoël. **Italian translation**: 1979. *Cronache della mia vita*. Milan: Feltrinelli.

TASSO, Torquato. 1853. *Lettere* (C. Guasti ed.). Firenze: Le Monnier.

———. 1875. *Prose diverse nuovamente raccolte ed emendate* (Cezare Guasti ed.), VOL. 1. Firenze: Le Monnier.

———. 1958. *Dialoghi* (E. Raimondi ed.), VOL. 1. Firenze: Sansoni.

———. 1959. *Prose* (E. Mazzali ed.). Milan: Ricciardi.

TUAN, Yi-Fu. 1974. 'Space and Place: Humanistic Perspective'. *Progress in Geography* 3(6): 233–51.

TOBLER, Waldo. 1963. 'Geographic Ordering of Information'. *The Canadian Geographer* 2(4): 203–05.

TONIZZI, Elisabetta M. 1999. *Le grandi correnti migratorie del '900*. Turin: Paravia.

ULLMANN, E. L. 1953. 'Human Geography and Area Research'. *Annals of the Association of American Geographers* 43: 54–66.

VAGNETTI, Luigi. 1980. 'Mieux vaut voir que courir 1' in Giulio Macchi (ed.), *Cartes et figures de la Terre*. Paris: Centre Georges Pompidou, pp. 248–51.

VARELA, Consuelo (ed.). 1982. *Cristóbal Colón. Textos y documentos completes.* Madrid: Alianza Editorial. **Italian translation**: 1992. *Cristoforo Colombo. Gli scritti*, Turin: Einaudi.

VARELA, Francisco. 1983. 'L'auto-organisation: de l'apparence au mécanisme' in Paul Dumouchel and Jean-Pierre Dupuy (eds), *L'auto-organisation. De la Physique au Politique.* Paris: Seuil, pp. 147–64.

VAYSSIÈRE, Bruno-Henri. 1980. 'La Carte de France' in Giulio Macchi (ed.), *Cartes et figures de la Terre.* Paris: Centre Georges Pompidou, pp. 252–65.

VELTMAN, Kim H. 1980. 'Ptolemy and the Origins of the Linear Perspective' in Marisa Dalai Emiliani (ed.), *Atti del convegno internazionale di studi: la prospettiva rinascimentale.* Florence: Centro Di, pp. 403–07.

VERNANT, Jean-Pierre. 1962. *Les origines de la pensée grecque.* Paris: Presses Universitaires de France.

———. 1966. 'Mythe et pensée chez les Grecs. Études de psychologie historique'. *Annales. Économies, Sociétés, Civilisations* 21(6): 1305–08. **Italian translation**: 1978. *Mito e pensiero presso i Greci. Studi di psicologia storica.* Turin: Einaudi.

VIDAL DE LA BLACHE, Paul. 1904. 'La Carte de France au 50.000'. *Annales de Géographie* 13: 112–20.

———. 1911. 'Les genres de vie dans la géographie humaine'. *Annales de Géographie* 20: 193–212, 289–304.

———. 1913. 'Des caractères distinctifs de la Géographie'. *Annales de Géographie* 23: 289–99.

———. 1922. *Principes de géographie humaine.* Paris: Colin.

———. 1926. *Principles of Human Geography* (Milicent Todd Bingham trans.). London: Constable.

VOLK, Tyler. 1995. *Metapatterns: Across Space, Time, and Mind.* New York: Columbia University Press.

WEBBER, Melvin M. 1964. 'The Urban Place and the Nonplace Urban Realm' in Melvin M. Weber (ed.), *Explorations into Urban Social Structure.* Philadelphia: University of Pennsylvania Press, pp. 47–68.

WEBER, Max. 1951. *Gesammelte Aufsätze zur Wissenschaftslehre.* Mohr: Tübingen. **Italian translation**: 1958. *Il metodo delle scienze storico-sociali*, Turin: Einaudi.

WEYL, Hermann. 1952. *Symmetry.* Princeton, NJ: Princeton University Press. **Italian translation**: 1962. *La simmetria.* Milan Feltrinelli.

WHEATLEY, Paul. 1971, *The Pivot of the Four Quarters: A Preliminary Enquiry into the Origins and the Character of the Ancient Chinese City*. Edinburgh: Edinburgh University Press.

WITTGENSTEIN, Ludwig. 1922. *Tractatus Logico-philosophicus*. London: Routledge and Kegan Paul. **Italian translation**: 1989. *Tractatus logico-philosophicus*. Turin: Einaudi.

——. 1956. *Bemerkungen über die Grundlagen der Mathematik*. Frankfurt am Main: Suhrkamp. **Italian translation**: 1971. *Osservazioni sopra i fondamenti della matematica* Turin: Einaudi.

——. 1969. *On Certainty* (Denis Paul and G. E. M. Anscombe trans). Oxford: Blackwell.

——. 2010. *Remarks on Frazer's 'Golden Bough'* (Rush Rhees ed., A. C. Miles trans.). Corbridge: Brynmill Press.

WUNENBURGER, Jean-Jacques. 1997. *Philosophie des images*. Paris: Presses Universitaires de France. Italian translation: 1999. *Filosofia delle immagini*. Turin: Einaudi.

ZELLINI, Paolo. 1999. *Gnomon. Un'indagine sul numero*. Milan: Adelphi.

ZEVI, Bruno. 1960. *Biagio Rossetti architetto ferrarese: il primo urbanista moderno europeo*. Turin: Einaudi.